United States Nuclear Regulatory Commission

Protecting People and the Environment

NUREG-1959

I0482630

Intrusion Detection Systems and Subsystems

Technical Information for NRC Licensees

Office of Nuclear Security and Incident Response

AVAILABILITY OF REFERENCE MATERIALS
IN NRC PUBLICATIONS

United States Nuclear Regulatory Commission

Protecting People and the Environment

NUREG-1959

Intrusion Detection Systems and Subsystems

Technical Information for NRC Licensees

Manuscript Completed: November 2010
Date Published: March 2011

Prepared by:
Office of Nuclear Security and Incident Response

Office of Nuclear Security and Incident Response

ABSTRACT

This report provides information relative to designing, installing, testing, maintaining, and monitoring intrusion detection systems (IDSs) and subsystems used for the protection of facilities licensed by the U.S. Nuclear Regulatory Commission. It contains information on the application, use, function, installation, maintenance, and testing parameters for internal and external IDSs and subsystems, including information on communication media, assessment procedures, and monitoring. This information is intended to assist licensees in designing, installing, employing, and maintaining IDSs at their facilities.

CONTENTS

LIST OF FIGURES

LIST OF TABLES

1. INTRODUCTION

This report provides information about designing, installing, testing, maintaining, and monitoring intrusion detection systems (IDSs) and subsystems used for the protection of facilities licensed by the U.S. Nuclear Regulatory Commission (NRC). Title 10 of the *Code of Federal Regulations* Part 73, "Physical Protection of Plants and Materials," addresses the NRC's requirements for the physical protection of nuclear power reactor facilities, independent spent fuel storage facilities, fuel cycle facilities, strategic special nuclear materials, and special nuclear material of moderate and low strategic significance. The specific requirements for physical protection programs at NRC-licensed facilities are determined by the type of NRC licensee and the special nuclear material authorized to be possessed by that licensee.

The use of IDSs at NRC-licensed facilities is one component of a physical protection program, the purpose of which is to ensure that only authorized personnel gain access to areas containing nuclear materials and that the use of nuclear materials does not endanger the public health and safety.

It is important to note that the effectiveness of a physical protection program depends on the coordination and integration of all the various components of that program, from equipment, to policies, to personnel, as well as the ability of the licensee to monitor and maintain these components to ensure they operate as intended. The use of IDSs is just one component of the overall physical protection program. Therefore, each licensee should design, install, and maintain an IDS that complements the needs and implementation of the other components of the physical protection program.

The following sections of this report provide information pertinent to the application, use, function, installation, maintenance, and testing parameters for internal and external IDSs and subsystems, including information on alarm communication media, assessment technology, and monitoring. The information is intended to assist licensees in designing, installing, employing, and maintaining IDSs at their facilities. When applying this information, it is important to confirm that site-specific conditions are analyzed and accounted for and that manufacturer recommendations for system application, installation, maintenance, and testing are followed to ensure the system capabilities are consistent with the physical protection program goals and that the systems are capable of performing their intended functions.

2. INTRUSION DETECTION SYSTEMS

2.1 Overview

Intrusion detection and assessment systems are an integral part of any physical protection system. Detection and assessment provide a basis for the initiation of an effective security response. Intrusion detection systems (IDSs) should be designed to facilitate the detection of attempted and actual unauthorized entry into designated areas and should complement the security response by providing the security force with prompt notification of the detected activity from which an assessment can be made and a response initiated. The method(s) of detection and assessment selected for implementation should be robust and be capable of providing the highest level of protection for the specific application. In the design of the detection and assessment aspects of a physical protection system, various methodologies should be incorporated to provide a fully integrated detection capability. The integration of various detection and assessment methodologies not only contributes to a superior detection and assessment capability, but also provides multiple overlapping layers that support each other if one method fails.

2.2 Intrusion Detection Equipment

2.2.1 Premise Control Unit

A premise control unit (PCU) is a device that receives changes of alarm status from sensors and transmits an alarm condition to the alarm monitoring station. The PCU also allows authorized personnel to change the alarm zone status of the alarm zone.

To be useful, an intrusion alarm system should have the capability to restrict status changes to only those who possess proper authority such as alarm system administrators or alarm station operators. A system may restrict access to performing system functions in any number of ways, often with keys or codes used at the PCU or a remote PCU near an entry. High-security alarms may require multiple codes, a fingerprint, badge, hand-geometry, retinal scan, encrypted response generator, and/or other means that are deemed sufficiently secure for the purpose.

To function as intended, the PCU must be located within the secured area to protect it from unauthorized access. The alarm status is normally delayed (for no more than 30 seconds) to allow the entrance authorization process to be completed. Like the sensors, the PCU requires a primary power source for operation and should be connected to an emergency power source to remain operable during the loss of primary power. The PCU should be equipped with a tamper alarm to detect attempts at unauthorized manipulation. All changes in status of the alarm zone must be transmitted to the alarm monitoring stations. Examples of reportable status include but are not limited to access/secure, tamper, loss of power, improper access code or procedure, test or maintenance mode, and masked sensors. PCUs can be configured with duress capabilities, which will allow personnel to surreptitiously notify the alarm monitoring station of a compromised situation.

2.2.2 Transmission and Annunciation

To ensure transmission line security, when the transmission line leaves the facility and traverses an uncontrolled area, Class I or Class II line supervision should be used.

2.2.2.1 Class I

Class I transmission line security is the security achieved through the use of a data encryption system or an algorithm based on the cipher feedback or cipher block chaining mode of encryption. To ensure Class I transmission line security, the data encryption system used must be certified by the National Institute of Standards and Technology or another independent testing laboratory.

2.2.2.2 Class II

Class II transmission line security refers to systems in which the transmission is based on pseudo random-generated tones or digital encoding using an interrogation and response scheme throughout the entire communication, or Underwriters Laboratories Class AA line supervision. The signal must not repeat itself within a minimum 6-month period. Class II transmission line security must also be protected against compromise using resistance, voltage, current, or signal substitution techniques.

Note: Direct current (dc) line supervision should not be used because it is highly susceptible to manipulation or substitution to circumvent valid alarm annunciation. Additional information on alarm communication and display (AC&D) can be found in Section 8 of this report.

2.2.3 Intrusion Detection System Cabling

Cabling designs that combine signal transmission lines and power source cabling into one of the two alarm monitoring stations or one common area before they reach the alarm monitoring stations should be avoided. Facilities should, to the extent practicable, ensure that the cabling (signal transmission and power) to each alarm station is sufficiently separated and not accessible to external manipulation. The separation and nonaccessibility of cabling provide a level of assurance that at least one alarm station will maintain the capability to perform required alarm station functions if the cabling is subject to an occurrence that results in the loss of functionality within the other alarm station. Key considerations in the design and installation of alarm system cabling are the prevention of unauthorized and external manipulation and minimization of the exposure to environmental conditions. It is important to identify and evaluate intrusion detection and assessment system cabling that may be vulnerable to disruption by external manipulation (e.g. adversarial action) and consider further protective measures such as hardening or periodic surveillance.

The cabling between the sensors and the PCU should be dedicated to the intrusion detection equipment (IDE) for which it provides service and should comply with national and local code standards. Cabling should be installed in metal conduit between the alarmed area(s) (i.e., controlled access area (CAA), vital area (VA), or material access area (MAA)) PCU and the alarm monitoring stations. Exceptions may apply when the cabling is not easily accessible, such as when it is run underground, traverses or is within another controlled area, or is encrypted with Class I or II encryption.

2.2.4 Entry Control Systems

If an entry control system is integrated into an IDS, reports received in the alarm stations from the automated entry control system should be subordinate in priority to reports from intrusion alarms. Additional information on AC&D appears in Section 8, "Alarm Communication and Display," of this report.

2.2.5 Power Supplies

Primary power sources for IDE may be commercial alternating current (ac) or dc power. In the event of a failure of primary power, the system should be designed to provide an indication that the IDS is operating on secondary, backup, or emergency power sources without causing false alarm indications. Section 9 of this report presents additional information on backup and emergency power.

2.2.5.1 Emergency Power

Emergency power should enable IDE to remain operable during the loss of primary/normal power. An emergency power supply should consist of a protected independent power source that provides a minimum of 4 hours operating power battery and/or generator power. When batteries are used for emergency power, they should be maintained at full charge by automatic charging circuits. The manufacturer's periodic maintenance schedule should be followed and results documented.

2.2.5.2 Power Source and Failure Indication

A visual (illuminated) and audible indication of power failure should be provided at the PCU of the power source in use (ac or dc) and in the alarm monitoring stations. Equipment at the alarm stations should indicate a failure in power source, a change in power source, and the location of the failure or change.

2.3 System Considerations

2.3.1 Maintenance Mode

When an alarm zone is placed in the maintenance mode, this operability status should be signaled automatically to the alarm monitoring stations. The signal should appear as an alarm or maintenance message at the alarm stations, and the IDS should not be securable while in the maintenance mode. The alarm or message should be visible at the alarm stations continually, throughout the period of maintenance. A standard operating procedure should be established to address appropriate actions when maintenance access is indicated at the panel. A record of each maintenance period should be archived electronically in the system or on other electronic media, or it may be printed in hard copy, in accordance with site procedures. A self-test feature should be limited to one second per occurrence. Section 8 "Alarm Communication and Display," of this report presents additional information on AC&D.

2.3.2 Shunting or Masking Condition

Shunting or masking of any alarm zones or sensors should be appropriately logged or recorded in the archive. A shunted or masked alarm zone or sensor should be displayed as such at the alarm monitoring stations throughout the period that the condition exists. Section 8 "Alarm Communication and Display," of this report presents additional information on AC&D.

2.3.3 Component Tamper Protection

IDE components, to include all sensors, control boxes, and junction boxes, should be equipped with tamper protection and indication to provide deterrence, delay, and detection of attempted unauthorized manipulation.

2.3.4 System Component Status Changes

No capability should exist to allow changing the status of the IDE and associated access control equipment from a location outside the area for which the equipment is applied. If used, PCUs should be located within a secure area that is equipped with intrusion detection capabilities that annunciate in the alarm monitoring stations. Only alarm station operators or authorized technicians should initiate status changes for IDE and associated access control equipment. Operation of the PCU should be restricted by use of a device, password, or procedure that verifies the authorization of an individual to make such changes. For applications of increased security, status changes for IDE and associated access control equipment may require confirmation from an additional system operating authority before the change is allowed to occur (e.g., a concurrence action from an alarm station operator confirming the status change initiated by the other alarm station operator or authorized technician).

2.3.5 False and/or Nuisance Alarm

A false alarm is an alarm generated without apparent cause. False alarms are generally attributed to electronic phenomena, whether of the sensor itself or of the electrical infrastructure of the sensor system. Alarm signals transmitted in the absence of detected intrusion, or when the cause cannot be verified as a nuisance alarm, are false alarms. A nuisance alarm is the activation of an alarm sensor by some influence for which the sensor was designed but which is not related to an intrusion attempt. To validate detection, all alarms must be assessed and may require further investigation to determine the cause. The results of all alarms should be documented. The design goal of an intrusion detection system should be to limit false alarms and nuisance alarms to a total of not more than one false alarm per zone per day and one nuisance alarm per zone per day. A sound maintenance program is often a key factor in minimizing the number of false and nuisance alarms an IDS system generates.

Alarm records should be maintained for effective preventive maintenance. These documented records support effective preventive maintenance. These records should include a daily nuisance alarm log. A daily nuisance alarm log can aid in the tracking of equipment or sectors where maintenance is required. This alarm log should contain an assessment of the cause of the alarm and should be referenced by alarm sector, specific location within the sector, and, as applicable, by component serial number, which provides unit-specific tracking. The alarm log enables problems related to a specific alarm sector to be investigated and identified for corrective action.

2.3.6 Installation, Maintenance, and Testing

Proper system installation, maintenance, and testing contribute to acceptable system performance and minimize its vulnerability to defeat. The defeat vulnerability of the sensor is an inherent characteristic of the sensor technology, how it is installed, and how sensors are configured within the system. For an effective detection system, intrusion detection sensors must be selected and properly installed to: maximize the detection probability, minimize its nuisance alarm rate, and minimize its vulnerability to defeat. Intrusion detection and assessment systems and components should be installed, maintained, and tested in accordance with manufacturers' specifications which account for system design, application, and performance. For IDS systems that are integrated with other components of the physical protection system, the specific configuration of all integrated components should be evaluated to ensure the system performs its intended function and provides the desired level of protection. Prior to installation the planned configuration of perimeter intrusion detection and assessment systems, physical barriers at the protected area perimeter (to include any associated nuisance barriers), and the isolation zones should be evaluated to ensure that the planned perimeter system design will satisfy the detection and assessment goals of the physical protection program. After installation, the configuration of these components should be re-evaluated based on the zone of detection provided by the IDS when operating to ensure that the IDS provides the required coverage and eliminates the potential for unauthorized bypass. The configuration of the components of the protected area perimeter (intrusion detection and assessment systems, physical barriers and the isolation zones) when combined with the IDS zone of detection should account for common defeat methodologies such as walking, crawling, running, climbing, jumping and bridging consistent with the design goals of the physical protection program. As such, perimeter IDS testing should include all defeat methodologies that could be used to defeat the system. With consideration for safety, simulations of certain defeat methodologies (e.g. jumping, bridging) during testing may be necessary.

Intrusion detection and assessment system or component installation, maintenance, and testing should be performed only by trained and qualified personnel. All installation, system settings/sensitivity levels, maintenance, and testing activities should be documented so that they may be used to support system or component troubleshooting and to ensure proper performance, maintenance, and testing intervals.

To ensure that a system or component continues to provide the level of protection intended, maintenance, testing, and calibration activities should be performed at intervals consistent with manufacturers' specifications and the operational demand placed on the system or component. Most system and component manufacturers have established similar testing criteria, especially in the area of testing frequency. The information below reflects general manufacturer recommendations regarding intrusion detection and assessment system testing intervals.

2.3.6.1 Acceptance Testing

When a sensor is first installed, it must be tested in order to formally "accept" the sensor as a part of the physical protection system. Acceptance testing consists of two parts:

(1) A physical inspection to assure that the sensor was installed properly

(2) A performance test to establish and document the level of performance

2.3.6.2 Performance Testing

Performance testing should be conducted every 6 months or whenever a system or component is returned to service after modification, repair, maintenance, or an inoperable state. The performance test should include the use of the documented levels of performance from the original acceptance testing to verify that the system or component is still performing adequately.

2.3.6.3 Operability Testing

For systems or components in continuous operation, operability tests should be performed weekly to confirm that the sensors are operational and that alarms are communicated to the alarm station display system. These tests should be scheduled such that all sensors are tested at least once every 7 days. It is recommended that testing be performed on a portion of the total sensors over a distributed time period so that some sensors are tested during different times of the day within the 7-day timeframe. Subsequent testing should ensure that each sensor is tested during a different time of day than the previous test conducted. When conducting perimeter IDS operability testing in conjunction with perimeter assessment equipment testing, varying the time of day a sensor is tested enables the verification of assessment capabilities during the varied illumination conditions of daylight and at night.

3. EXTERIOR DETECTION SENSORS

3.1 Microwave Sensors

3.1.1 Principles of Operation and Performance

Microwave sensors are motion detection devices that flood a specific area with an electronic field. A movement in the area disturbs the field and generates an alarm. Microwave sensors are typically used for detection in long, flat, narrow perimeter zones. (See Figure 1.) They are classified as an active, visible, volumetric, freestanding, and line-of-sight sensor.

Figure 1: This photo depicts the application of microwave sensors within a perimeter intrusion detection and assessment system (PIDAS). Note that the drawn-in detection zone goes from the front microwave to another that is approximately 90+ meters (300+ feet) away. The middle microwave transmits to a unit behind the photographer. This configuration is referred to as a "basketweave" layout.

Microwave sensors transmit microwave signals in the "X" band. These signals are generated by a diode operating within preset limits that do not affect humans or the operation of pacemakers.

The effective detection zone width and height of a given microwave system will largely depend on the mounting height of the antennas when operating over the same type of surface at the same range. The mounting height is measured from the center of the antenna aperture to the ground. When properly aligned, the maximum detection height and width will occur at midrange. During installation in the field, antenna heights can be adjusted for maximum signal while monitoring the signal at the receiver. Most manufacturers indicate the linear operating range to be approximately 100–150 meters (328–492 feet) when detection of a stomach-crawling intruder is a requirement. Detection zone sizes are normally stated for midrange between the transmitter and receiver at the maximum antenna separation.

3.1.2 Types of Microwave Sensors

Microwave systems are found in two basic configurations:

(1)   bistatic, consisting of a transmitter and receiver antenna located at either end of a perimeter sector

(2)   monostatic, which uses a transceiver

The bistatic microwave link consists of a transmitter at one end of a long, flat location and a receiver module at the other end. The transmitter emits a modulated, low-power signal in the microwave frequency band. The signal at the receiver antenna is combined in a vector summation of the direct line-of-sight signal and the reflected signals from the ground and nearby objects. The receiver monitors this signal for small changes that occur when objects move in the detection zone of the sensor. When the changes in the zone exceed an established threshold, an alarm is generated.

Automatic gain control circuitry is used to allow the microwave sensor to compensate for very slow changes in signal because of environmental conditions.

The size and shape of the detection zone are determined by the antennas, the transmitted frequency, and the distance between the antennas. The detection zone typically resembles an oblate spheroid, much wider and taller in the middle of the detection zone and narrowing on each end. (See Figure 2.)

Figure 2: This diagram depicts the detection zone of a typical bistatic microwave.

A monostatic microwave unit consists of a transmitter and receiver located in the same unit. (See Figure 3.) The two types of monostatic microwaves are amplitude modulated (AM) and frequency modulated (FM). AM monostatic microwave systems detect changes in the net vector summation of the received signal, similar to a bistatic system. FM monostatic systems operate on a pulsed Doppler principle and thus can determine range information. Some pulsed Doppler microwave sensors can be set up to only detect motion moving toward or away from the sensor in order to help minimize nuisance alarms. These are finding usage in some specialty applications, such as on ladders or stairs where the adversary has only limited approach options.

In general, the useful range of a monostatic microwave is considerably less than that of a bistatic system. Monostatic microwave sensors are usually recommended only for application in areas inappropriate for a bistatic microwave.

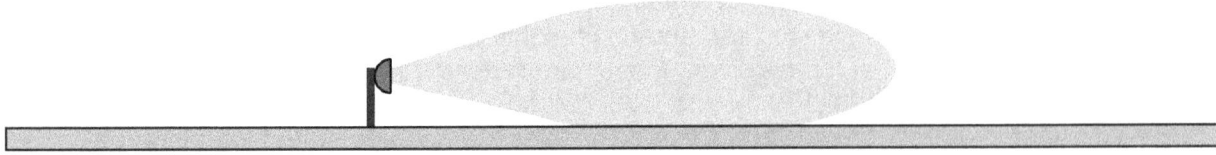

Figure 3: This diagram depicts the detection zone of a typical monostatic microwave.

3.1.3 Sources of Nuisance Alarms

While a microwave works fairly well in most weather conditions, some types of weather extremes may cause nuisance alarms, in particular, the first few seconds of a hard rain. Normally, after the rain has started, the sensor recognizes and adapts to the rain. Large puddles of standing water can cause nuisance alarms if wind creates ripples in its surface. Running water is also a major source of nuisance alarms if proper drainage is not designed into the sensor zone.

Most flying debris, such as leaves, paper, cardboard, and tumbleweeds, will not cause an alarm because these objects are not reflective. However, if these same objects were wet or icy, they would likely cause the system to alarm. An aluminum can, because it is highly reflective, will also cause an alarm.

While an animal as small as a mouse or a single bird will not normally be detected by a microwave, animals the size of a rabbit or a flock of birds likely will be detected.

If the fabric of a chain-link fence moves, the reflection may cause an alarm, even though the fence is outside of the volume of detection.

3.1.4 Characteristics and Applications

The following list describes characteristics and applications of microwave sensors:

- The microwave is a volumetric sensor and has a large detection pattern.

- To a potential adversary, the detection pattern of the microwave is unknown.

- Microwaves may be stacked to create a much taller detection pattern.

- The microwave works well in most weather conditions.

- The microwave is appropriate for long, narrow, flat zones.

- The microwave will be able to detect within snow, as long as the antenna is not blocked.

- Microwaves are typically vulnerable to crawlers immediately in front of both the transmitter and receiver.

- Microwave sensors do not work well in extremely heavy fog or in a wet, heavy snow.

- Microwaves will not perform well in a rainy climate when drainage is poor.

- Microwaves perform less optimally in areas of heavy snow.

- Microwave sensors do not lend themselves to use in areas where the terrain has large variations.

- Site preparation for a microwave installation can be difficult because of the requirement for very flat terrain.

- Microwaves will not work well in a zone between two fences that are not at least 6 meters (20 feet) apart.

- Vulnerabilities have been noticed when microwaves are operated near runways at airports.

3.1.5 Installation Criteria

Typically, a bistatic microwave perimeter detection system should be installed to operate effectively at a distance between antennas of not more than 120 meters (approximately 328 feet). Successive microwave links and corners should overlap to eliminate dead spots (areas where the microwave beam cannot detect) below and immediately in front of transmitters and receivers. The required amount of overlap of successive links is contingent on the antenna pattern, alignment, and mounting height but may be up to 10 meters (approx. 32.8 feet) in length for some models. Overlap should be adequate to provide continuous detection of crawling intruders over the entire sector. (Refer to Figure 4 and Figure 5.)

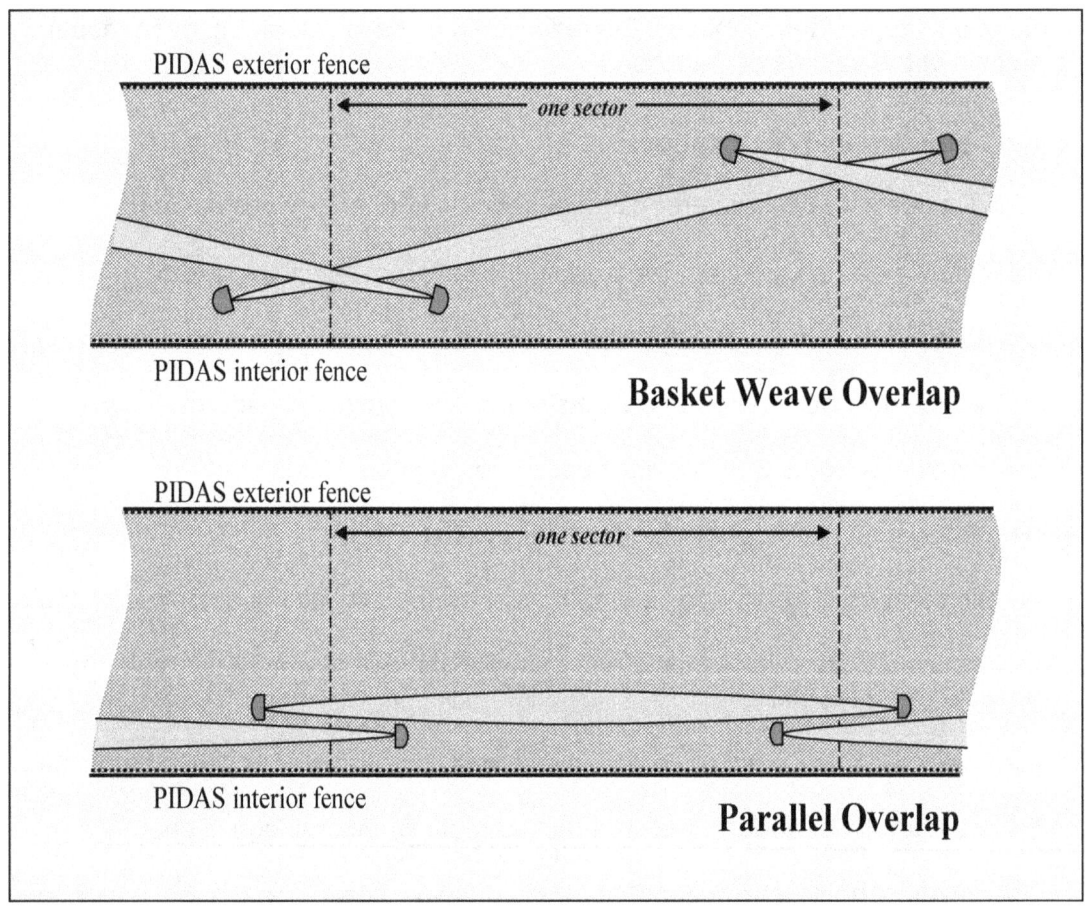

Figure 4: Examples of possible microwave sensor layouts.

Figure 5: Example of a corner setup of microwave sensors that provides overlap to decrease the possibility of an adversary crawling beneath the beams.

There should also be an area at each end of the sector where the crawler is detected by both microwave links to ensure continuous detection between adjacent sectors.

Since the bistatic transmitter/receiver link is a line-of-sight system, variations in ground level (e.g., ditches and valleys) could allow some intruders to crawl under the beam; bumps or obstructions may create shadows in the beam that the adversary could crawl behind; and variable obstructions (e.g., snow drifts or accumulations) may interrupt the beam. To prevent passage under the microwave beam, variations in the ground should be leveled, ditches should be filled, and obstructions should be removed so that the area between the transmitter and receiver is clear of obstructions and free of rises or depressions. Overlaps should be adequate to allow the microwave beam to expand enough to detect an intruder attempting to jump over the microwave beam.

In zone overlap areas, the equipment for the overlapped zones should usually both be transmitters, or both should be receivers to minimize the interference between the successive links. Different modulation frequencies are used for microwave links in adjacent sectors to prevent interference from other microwave transmitters.

Because standing water can be a major source of nuisance alarms, proper drainage is required to prevent standing water and to prevent erosion that may provide the adversary with low spots where crawling is more difficult to detect. Gravel is often used for the perimeter surface to allow water to drain quickly and minimize erosion. Solid surfaces, such as asphalt or concrete, may prevent erosion problems but may actually cause a higher nuisance alarm rate during periods of heavy rain, because of the rain bouncing off of the surface.

Each microwave unit should be mounted securely on rigid posts and aligned according to the manufacturer's installation criteria. The mounting pole of the microwave for the adjacent sector should not provide an aid to the adversary attempting to jump over the microwave beam. (Refer to Figure 6.) Cutting down the mounting pole or using a spike or other attachment that does not allow the adversary a solid jumping platform may be required.

Figure 6: This scene taken from a test of microwave sensors shows the possibility of an attacker using the microwave structure as an aid to jump over its own detection zone. Note: The microwave patterns shown in the photo were added for illustration purposes.

Because of variances in the antenna patterns of different microwave systems, the height may need to be varied slightly to obtain coverage adequate to detect crawling intruders. Accordingly, the mounting mechanisms for a system should permit adjustment of antenna height and position to correct poor performance or alignment.

Wiring should be protected with conduit, using flexible conduit near the sensor to allow alignment and adjustment of the antenna.

The clear area (the space between the two PIDAS fences) should be sufficiently wide to preclude the generation of alarms by legitimate movements near the microwave link (e.g., personnel walking or vehicular traffic) and to preclude system degradation caused by reflections from any structure, such as the perimeter fence. The manufacturer should provide approximate dimensions of the microwave pattern. Since the beam is relatively wide, care should be taken to ensure that reflections from authorized activities do not create nuisance alarms. With the microwave link installed inside a perimeter barrier or between a double perimeter barrier, the transmitter and receiver should be positioned so the height of the zone of detection will detect anyone jumping or attempting to bridge over the microwave beam into the protected area from atop perimeter barriers such as fences or walls. Typically, the distance between a chain-link security fence with an overall height of 2.4 meters (8 feet) and the center of the microwave beam should be a minimum of 2.4 meters (8 feet). In addition, the microwave link should be positioned within the isolation zone to maximize the zone of detection and enhance assessment once detection is made. Neither a transmitter nor a receiver should be mounted on a fence.

Stacking of microwave sensors is one means of increasing the detection zone height of the system to enhance its detection capabilities. The stacking technique places a microwave link close to the ground to provide better crawl detection and one or more links higher to improve detection of jumping or bridging. (Refer to Figure 7.) The stacked units may use different carrier frequencies, different antenna polarization, or multiplexing to prevent interference between the microwave links.

Figure 7: An example of stacked microwave sensors. Note that the possibility of crawling beneath the microwave detection zone has been reduced significantly because of the microwave sensor that is installed close to the ground.

3.1.6 Testing

A regular program of testing of sensors is imperative for maintaining them in optimal operating order. Three types of testing need to be performed during the life of a sensor: acceptance testing, performance testing, and operability testing.

3.1.6.1 Acceptance Testing

When a microwave sensor is first installed, it should be tested in order to formally "accept" the sensor as part of the physical protection system. Acceptance testing consists of two parts: a physical inspection and a performance test, as follows:

(1) A **physical inspection** to ensure that the microwave unit was installed properly consists of the following:

- Verify that the installation matches the installation drawings, which should follow the guidance provided by the manufacturer.
- Verify sector intersection spacing.
- Verify that signal and power wires are routed in the conduit.
- Verify proper power levels (voltage and amperage).
- Verify correct wire connections.

(2)   A **performance test** establishes and documents the level of performance. The following section describes the type of tests that are recommended for the microwave.

3.1.6.2   Performance Testing

Generally accepted performance criteria for a microwave detection system are that the system should be capable of detecting an intruder weighing a minimum of 35 kilograms (77 pounds) passing through the zone of detection between the transmitter and receiver, including the area in front of both the transmitter and receiver, whether the individual is walking, running, jumping, or crawling. Provisions should be made to ensure detection in spite of the dead spots in front of transmitters and receivers.

3.1.6.2.1   Procedure 1: Crawl Tests

Conduct 30 crawl tests within the detection zone/sector to verify that the probability of crawl detection is at least 90 percent at a 95-percent confidence level.

Each sector should be tested 30 times at points along the length of the microwave detection zone, though the overlap areas of two microwaves should be the primary focus.

As it can be difficult to locate an adult of approximately 35 kilograms (77 pounds) and because crawling is viewed as the most favorable form of spoofing a microwave sensor, most sites use a 30-centimeter (approx. 11.8 inches) diameter aluminum sphere to simulate crawl tests. (Refer to Figure 8.) The sphere diameter represents the cross-sectional area (at microwave frequencies) of an intruder lying parallel to the beam centerline. Pulling the sphere across the zone at slow speeds can determine if the microwave sensor is capable of detecting a small, stealthy individual. Because the radar cross-section of the sphere does not change with its orientation and the physical demands on the tester are much less than performing actual crawl tests, the sphere test can yield more repeatable results that eliminate variability because of the size and skill of actual crawling individuals.

Figure 8: An aluminum sphere is commonly used for testing the potential success of a crawler against a microwave detection system.

If the aluminum sphere is not detected until after it has been pulled across the centerline of the microwave detection zone, this often indicates an area of weak detection that may need to be addressed through additional adjustment of the sensor. A small individual can perform actual crawls in limited locations to supplement the sphere drag testing, perhaps testing in areas where the sphere was not detected until after it had crossed the beam centerline.

The aluminum sphere should be mounted on a nonmetallic platform or base that does not raise the height of the sphere or increase the radar cross-section. The base is usually a sled but is sometimes wheeled, which helps prevent the sphere from becoming dented. As the sphere begins to get dented, the radar cross-section increases and it no longer accurately represents a small individual. The cord used to pull the sphere across the zone should be nonmetallic so as not to affect the beam. Nylon sash cord can be used effectively. The cord should be long enough to allow the testers to be completely out of the microwave detection area. (Refer to Figure 9.)

Figure 9: Testing microwave sensors with an aluminum sphere.

3.1.6.2.2 Procedure 2: Other Tests

Another method of performance testing incorporates the use of a combination of the common defeat methodologies (walking, running, jumping, climbing, crawling and bridging) within each detection zone to verify a system's detection capability. Each sector should be tested 30 times at points along the length of the detection zone. Walk tests may be useful for mapping out a sensor's detection pattern for the purpose of identifying antenna misalignments, but the detection of a walker is not normally considered "challenging" for a microwave detection system.

Jumping, climbing and bridging tests should be performed to verify that the height and overlap of the beam is adequate. Jumps, assisted, unassisted and using nearby objects as a jumping platform should be performed to verify that detection is adequate. Run tests are suggested because some microwave sensors have a high-speed cutoff in order to eliminate some types of nuisance alarm sources. When using various defeat methodologies for performance testing, each defeat methodology should be used in the different areas of the zone being tested.

3.1.6.3 Operability Testing

The operability test may be conducted by walking into the expected detection zone of the microwave.

If no alarm is received, a maintenance request should be immediately generated and implementation of compensatory measures should be considered based on site-specific requirements.

3.1.7 Maintenance

Recommended maintenance procedures include the following:

- Vegetation or erosion in the areas of the microwave detection pattern should be checked.

- If the sensor has not been experiencing any recent problems (i.e., failure to alarm during a walk test) and if the most recent performance test was successful, there is then no reason to work on the alignment or sensitivity setting of the unit.

- At least one manufacturer recommends waxing the antenna covers periodically.

- Looking for any evidence of damage to the sensor or tampering with the device.

3.2 Electric Field Sensors

3.2.1 Principles of Operation

Electric field sensors (often referred to as e-field sensors) consist of posts or other supports with wires strung between them. (Refer to Figure 10.) In the sensor's most basic form, there is one field wire and one sense wire. An oscillating electrical signal is applied to the field wire, which establishes an electromagnetic field around the field wire. A sense wire is located close to and parallel to the field wire. Some of the field energy is coupled to the sense wire. A processor analyzes the received energy and monitors changes to the energy level. Most of the signal coupling results from capacitive coupling. The energy level coupled to the sense wire is dependent on the electrical ground (also referred to as earth ground), the spacing of the wires, the length of the wires, and the dielectric of the medium (air) between the wires. A human body is mostly water and the dielectric of water is about 100 times greater than that of air. As an intruder approaches the wires, the signal coupled to the sense wire is disturbed because of the higher dielectric of the intruder's body and because the intruder is partially in contact with an electrical ground provided by the earth. Figure 11 shows a typical electric field sensor's zone of detection.

Figure 10: Example of an electric field sensor installed in a test environment.

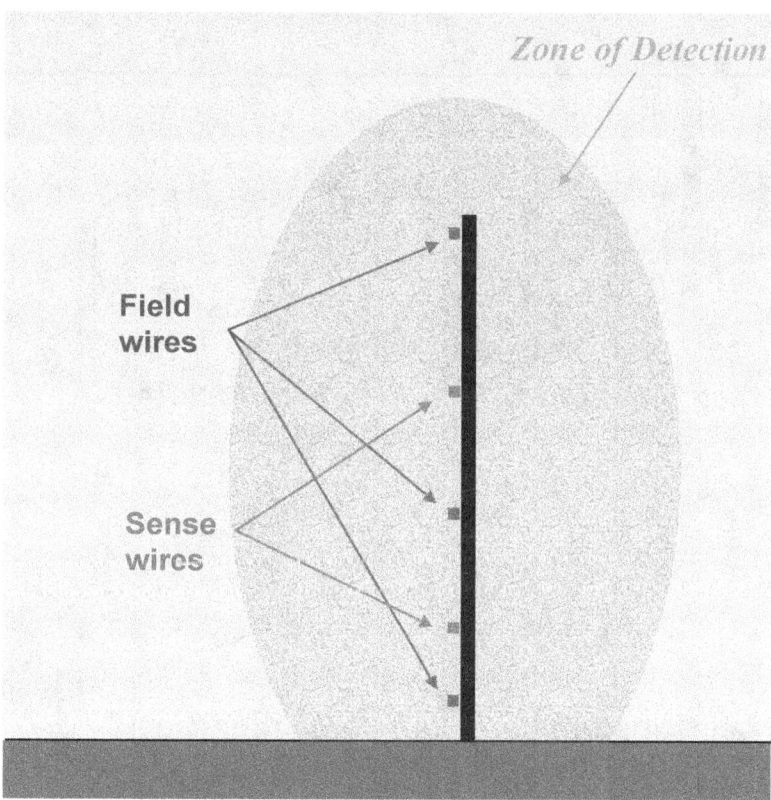

Figure 11: A cross-section of an electric field sensor's zone of detection.

Older electric field sensor processors were simply based on detecting a sufficient change in amplitude (threshold) of the analog signal that did not occur too fast or too slowly (bandpass filtering) to be considered an intruder. Unfortunately, many environmental sources could cause disturbances that would meet these simple conditions for an alarm. Newer models employ digital processing and sophisticated pattern recognition software to generate an alarm. This change has resulted in a significant reduction in nuisance alarms for new electric field sensors while maintaining good levels of detection.

Electric field sensors are volume detectors. They can usually follow the terrain where they are installed. While their characteristic wires give them the appearance of a line-type sensor, the electric field generated by these wires makes the zone of detection extend well beyond the wires. Electric field sensors can have a very wide detection volume (up to 2 meters or 6.5 feet wide) as well as a very tall detection volume (4.9 meters or 16 feet or more), based on the installation configuration. To achieve these heights, combinations of multiple field and sense wires can be incorporated within a single electric field system.

3.2.2 Types of Electric Field Sensor Applications

Electric field sensors can be installed in a freestanding configuration on their own posts. They can also be installed on chain-link fences, on the top of building roofs, on the side of building walls, and on the top of concrete walls if special supports are incorporated. (Refer to Figure 12.) Various wire configurations can be used to accommodate these mounting situations to yield the desired detection height and volume. Because of their adaptability to various vertical and horizontal surfaces, e-field sensors may be applicable to areas that are not compatible with the use of other sensors.

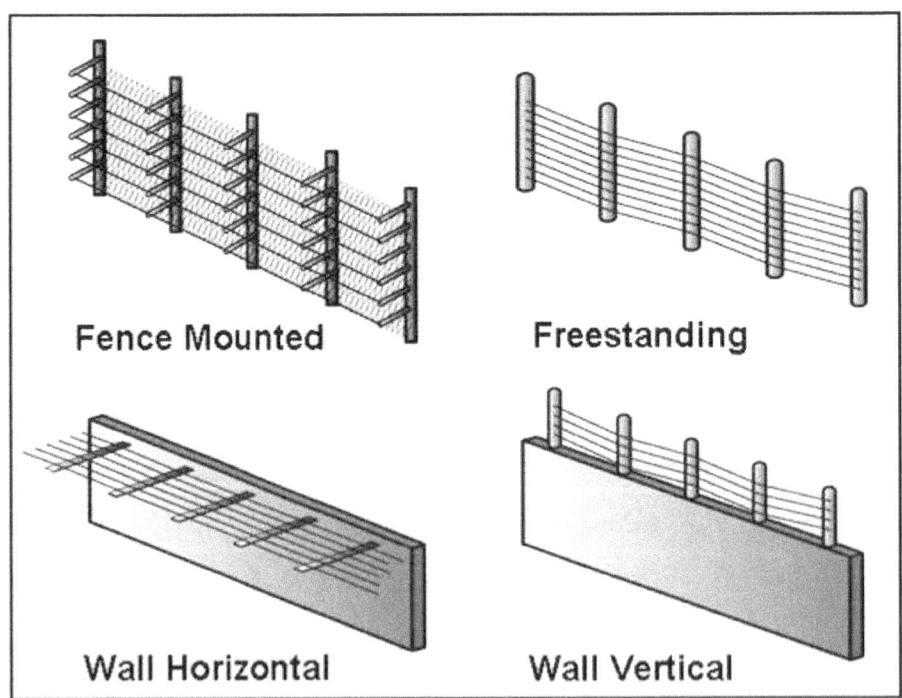

Figure 12: Various configurations for electric field sensors.

3.2.3 Sources of Nuisance Alarms

Proper installation of electric field sensors is critical to minimizing nuisance alarms. Any ungrounded metal object in the vicinity of an electric field sensor that can change potential can cause a nuisance alarm. For example, a chain-link gate may have grease in the hinges that

3-12

partially insulates it from the rest of the fence, but small wind-induced motion may occasionally result in contact between the gate and the metal of the hinge, which has the potential to cause a nuisance alarm. Attaching the sensor ground and all metal within the vicinity to a common electrical ground (earth ground) can reduce nuisance alarms.

Weather-related sources of nuisance alarms include rain, as well as wet or melting snow. During the start of a heavy rain, a few nuisance alarms should be expected. However, the newer electric field processors can quickly recognize the pattern of rain and, after recognition, will ignore it while still maintaining detection levels. During snowstorms, even with the bottom wire covered with snow, the sensor should still detect in an acceptable manner. During very heavy, wet snow, or when the snow is melting and possibly creating conduction paths to ground, there is a potential for nuisance alarms.

Small animals that touch the wires and larger animals that approach the wires may cause alarms. The use of nuisance control fences can significantly reduce these alarms. A single bird landing on a wire generally will not cause an alarm, but a flock of birds may. The wires of an electric field sensor are small enough that larger birds generally do not land on the wires.

Wet plants or wet weeds that touch the wires can also cause alarms. The use of nuisance control fences (to reduce blowing debris) and maintaining a sterile area (to prevent plant growth) near the wires can significantly reduce these nuisance alarms. Figure 13 shows an example of nuisance alarm sources.

Figure 13: Sources of nuisance alarms for electric field sensors include flocks of birds; wet, heavy snow; vegetation; and animals that may touch the wires.

3.2.4 Characteristics and Applications

The following list describes the characteristics and applications of e-field sensors:

- E-field sensors operate well in most environments.

- Configurations can be tuned to have a very large and relatively even detection volume, which can make an e-field sensor very difficult to defeat or circumvent.

- A variety of mounting configurations can be used (i.e., the sensor can be fence mounted on an existing chain-link fence (with some reduction in detection volume), or it can be mounted on its own posts in a freestanding configuration (preferably between two fences).

- E-field sensors can be installed on walls and roofs, which makes it easier to maintain a continuous line of detection if there is a building within the perimeter.

- E-field sensors can follow uneven terrain with posts spaced approximately every 6 meters (20 feet); the wires can be routed up and over gentle hills and valleys.

- The installation of e-field sensors usually requires a factory-trained installer or experienced alarm technician to perform installation.

- As the horizontal parallel wires of an e-field sensor naturally form a type of fence, this presents an operational barrier to the entry of authorized vehicles or personnel. Where personnel or vehicle access through the perimeter of a protected area is required, a different sensor (other than the e-field sensor) must be incorporated to allow for continuous and even detection through the entry area.

- Because this sensor is a volumetric sensor, large moving metal objects near the sensor can cause nuisance alarms. If the e-field sensor application includes an area where the frequency of vehicles passing close to the sensor could cause numerous alarms, a different sensor type should be considered.

- A potential defeat method for an e-field sensor is an intruder slowly attempting to crawl under the bottom wire. The bottom wire height should be evenly parallel to the ground and low enough to detect all crawlers. There should be no areas, such as a ditch caused by erosion or other low depression, where the wire height may create an area of vulnerability.

3.2.5 Installation Criteria

Many components are required for an e-field sensor installation. Figure 14 shows some of these parts, which include the sense wire, various insulators, posts, and support hardware.

Proper system installation is vital for proper performance; therefore, installers should be factory trained or experienced technicians and should follow manufacturer installation instructions.

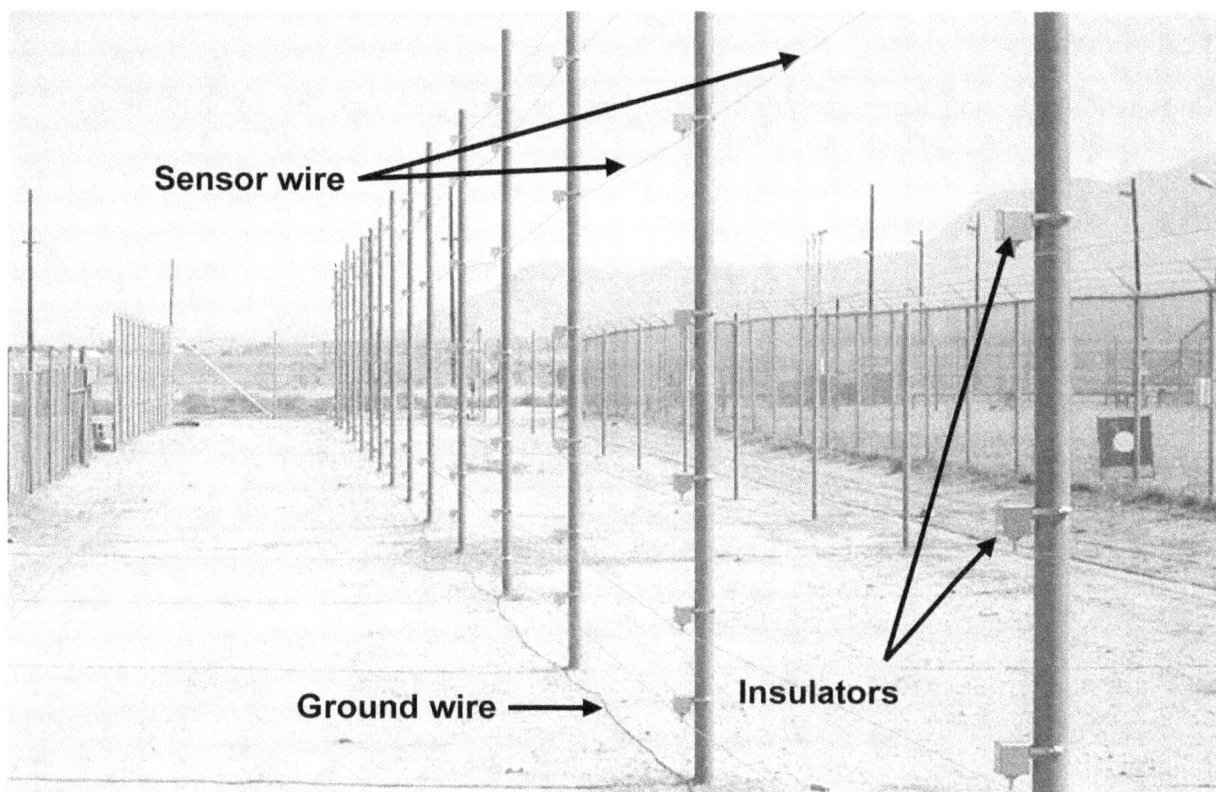

Figure 14: Main components of an electric field sensor installation.

The single most important detail in the installation process is that every piece of equipment or hardware must be connected to a common ground. This includes any electrically conductive structures or objects that are in the vicinity of the sensor system. The manufacturers of electric field sensors provide installation instructions, which specify grounding procedures that include the installation of a common ground wire and copper-clad steel rods that are driven into the ground.

Installed configurations may be freestanding (where the wires are mounted on their own dedicated poles) or may be attached to an existing sturdy fence. A freestanding installation is usually considered preferable; installation on a chain-link fence will effectively cut out half of the detection volume—that half on the side of the fencing that does not contain the electric field sense wires.

An electric field sensor system may also be mounted or attached directly to the wall or roof of a building. If the wall or building is not constructed of a material that can provide an adequate electrical ground (earth ground), a grounded metal screen can be installed under the sensor that will provide the needed ground reference.

A cleared area must be created where the sensor system will reside. Ideally, this should be an area where access is not possible by passing pedestrians or animals. Weeds near an electric field sense wire must be eliminated or kept well trimmed.

3.2.6 Testing

Three types of testing need to be performed during the life of a sensor: acceptance testing, performance testing, and operability testing.

3.2.6.1 Acceptance Testing

When a sensor is first installed, it should be tested before the formal "acceptance" of the sensor as part of the physical protection system. Acceptance testing consists of two parts:

(1)   A **physical inspection** to ensure that the sensor was installed properly, as follows:

- Verify that the installation matches the installation drawings, which should follow the guidance provided by the manufacturer.
- Verify sector intersection spacing.
- Verify that signal and power wires are routed in conduit.
- Verify proper power levels, both voltage and amperage.
- Verify correct wire connections.

(2)   A **performance test** to establish and document the level of performance (see next section).

3.2.6.2 Performance Testing

System or component modifications such as processor replacement or adjustment, wire spacing, or height adjustments, or any change to the sensor (or in the vicinity of the sensor) that might affect the detection capability would require performance testing to verify system performance.

The performance test should include a visual inspection of the sensor and of the general area where the sensor is installed. The test should record all relevant processor readings and compare these to the last recorded readings. It should measure the lower wire height at all the posts and midway between the posts and then compare these values to previous values documented when the sensor system was judged to have been performing optimally.

Because an electric field sensor is a volume detector, it will detect the approach of an intruder at some distance from the sense wires and an intruder penetrating the wire array. Though these systems or sectors of these systems may provide detection of an approach at some distance, in some cases the zone of detection may be such that detection occurs upon penetration of the zone between or in close proximity to the senor wires. The performance test must confirm that penetrations or attempted penetrations of the wire array that forms the electric field sensor system are detected at the required level. To ensure the overall effectiveness of an electric field system, performance testing should incorporate the use of a combination of common defeat methodologies (crawling, walking, jumping, climbing, and bridging) applicable to an electric field system within each detection zone. Each sector should be tested 30 times at points along the length of the electric field detection zone.

When using various defeat methodologies for performance testing, each defeat methodology should be used in the different areas of the zone being tested. If each of the 30 tests result in a detection, the electric field sensor sector has been confirmed to have at least a 90-percent probability of detection at a 95-percent confidence level. Particular attention should be given to any low depressions and to the areas where sectors overlap (one of the sectors must alarm in this area).

As an electric field sensor is most susceptible to defeat by crawling, during performance testing the 30 test attempts may include more crawl tests than the other defeat methodologies. For crawl tests the tester should lie down flat on the ground with his or her head toward the wires but starting outside the detection volume of the sensor. The tester should weigh 35 kilograms (77 pounds) or more. The tester should crawl slowly (approx. 2 to 6 inches per second or in accordance with manufacturers specifications) under the sense wires, keeping the motions as smooth as possible and stopping when a detection occurs (See Figure 15.). The crawler must fully penetrate the wire array (to a point beyond the detection volume on the opposite side of the sensor) to declare a nondetection.

Figure 15: An example of the crawl-test procedure.

For fence-mounted sensors, it is not possible to exit the detection volume because of the physical barrier of the fence. In this case, the sensor is not considered defeated simply by the tester's head or arm penetrating the plane of the wire array without a detection. The tester should continue to move their body through the wire array attempting complete penetration. Also, because an intruder must take some action to also penetrate the fence, such as cutting a hole in the fence or standing up to climb the fence, the intrusion test should include a reasonable simulation of the motion and tools required to simulate an intruder climbing over the fence, cutting through the fence, or going under the fence.

3.2.6.3 Operability Testing

Operability tests should be conducted by walking to the immediate vicinity of the electric field sensor wires and attempting to penetrate the zone of detection until an alarm is communicated and received in the alarm station.

During operability tests, the area in which the system or components are being tested should be surveyed for evidence of damage to equipment or possible tampering that may have occurred.

3.2.7 Maintenance

The routine maintenance required is mainly associated with inspecting the components and inspecting the near vicinity for ungrounded metal objects, plant growth, and/or debris. The sense wires of an electric field sensor are tensioned so that they vibrate at a frequency that is outside the signal frequency of interest. Verifying the tension of the sense wires and the inspection of system components should be part of the regular maintenance and may be combined with the scheduled testing of the sensor. An increase in nuisance alarms indicates the need for physical inspections.

Routine maintenance inspections should do the following:

- Inspect for weed growth, sterilize the soil or remove any growth near the sense wires.

- Inspect for and remove any debris that could cause nuisance alarms.

- Inspect the wire for proper tension and for any poor and/or broken connections.

- Inspect and clean dirt or debris (e.g. insects, spider webs, etc.) from insulators as needed.

- Inspect all mechanical parts for damage.

- Inspect all sensor electrical ground (earth ground) connections.

- Inspect nearby metallic objects for good ground connections.

- Confirm wire heights by looking for eroded or low areas where the wire height may be too high. Document the maximum bottom wire height determined to be acceptable to provide adequate detection.

3.3 Ported Coaxial Cable Systems

3.3.1 Principles of Operation

A ported coaxial cable system (commonly referred to as "ported coax") is a terrain-following, volumetric, covert intrusion detection system (IDS) that consists of two buried, ported coaxial cables and a processing unit. (Refer to Figure 16.) The processor contains a transmitter, a receiver, various amplifier and filter circuits, and a microprocessor with associated hardware and software. This system is an active electromagnetic sensor using two identical "leaky" coaxial cables (the shielding of the cable has holes or "ports" in it) buried parallel in the ground. The transmitter portion of the processor is connected to one cable, and the receiver is connected to the other.

Because the outer conductor of the cables is ported (i.e., it contains closely spaced small holes or gaps in the shield that allow radiofrequency (RF) energy to radiate), any electromagnetic energy injected into the transmitter cable is radiated into the surrounding medium, and some of this energy is coupled into the receiver cable through its ported shield. Thus, a static field of coupling is established between the cable pair. When an intruder enters the established field, the coupling is perturbed, and the change in received signal is digitally processed. Changes in this electromagnetic field that exceed threshold levels cause an alarm.

Figure 16: An illustration of the covert zone of detection for a ported coax sensor.

Ported coax sensors are sensitive to changes of dielectric or conductivity within the detection zone and are insensitive to seismic noise. The cross-section of the detection zone is somewhat elliptical and can be up to 1 meter (3 feet) high and 3 to 4 meters (9 to 12 feet) wide. Detection is achieved both above and, to some extent, below the ground. (Refer to Figure 17.) Tests have shown that both the detection zone size and detection sensitivity vary along the sensor cables. Such variance may result from several causes, such as phase cancellations, buried metallic objects, variation in cable separation or burial depth, and variable soil composition; therefore, it is important to identify the low-sensitivity areas and to adjust the sensitivity or the installation to achieve an adequate detection volume.

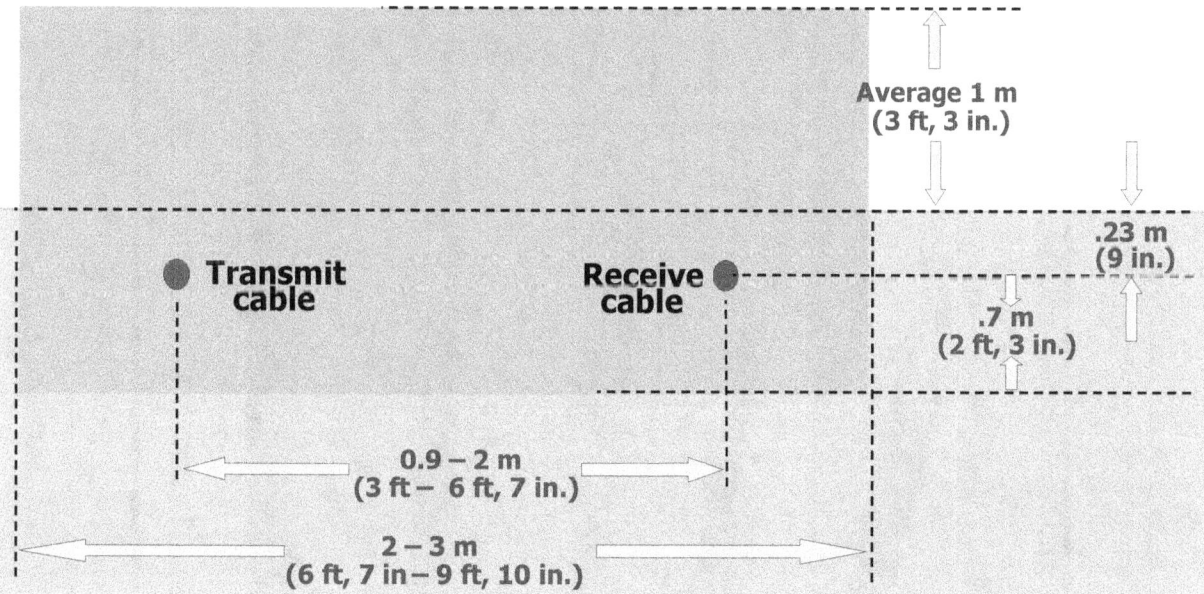

Figure 17: A typical installation and detection envelope (green oval) for a ported coax sensor.

Soil characteristics affect the signal strength of the ported coax receiver signal. However, even at sites where soil attenuation is relatively large, operational ported coax systems have been implemented. Variations in soil moisture are reflected in changes in ported coax signal attenuation. Wet soil is more conductive than dry or frozen soil; therefore, seasonal changes may be accompanied by changes in ported coax detection sensitivity. Recalibration is often required to account for seasonal changes.

3.3.2 Types of Ported Coaxial Cables Available

Either pulsed or continuous wave RF energy can be used in a ported coax sensor.

The pulsed system operates in principle as a guided radar, and therefore, both detection and location of the intruder (i.e., location along the cable length) are determined. Because of the ranging capabilities of this type of system, detection zones can be set in the software, enabling the configuration of more than the traditional two sectors per processor. This is possible because of the sensor's capability to divide each sensor cable into small subcells, usually 1 or 2 meters (3.28 or 6.56 feet) in length. Each of these subcells can be independently adapted to the site conditions and analyzed. This capability enables the alarm threshold to be varied in each of these subcells, providing uniform detection along the entire length of the cable.

Continuous wave ported coax sensors perform no range processing; therefore, they transmit continuously or during specified times. The frequency of operation is approximately 40 to 60 megahertz. There is only one alarm threshold per set of sensor cables, and the cables are the length of the sector. The continuous wave system detects the intruder but does not localize the presence along the cable length. Each processor can monitor one or two channels (or sectors) and provides an alarm indication if an intrusion occurs anywhere in the zone of detection.

Figure 18 shows examples of different ported coax cables.

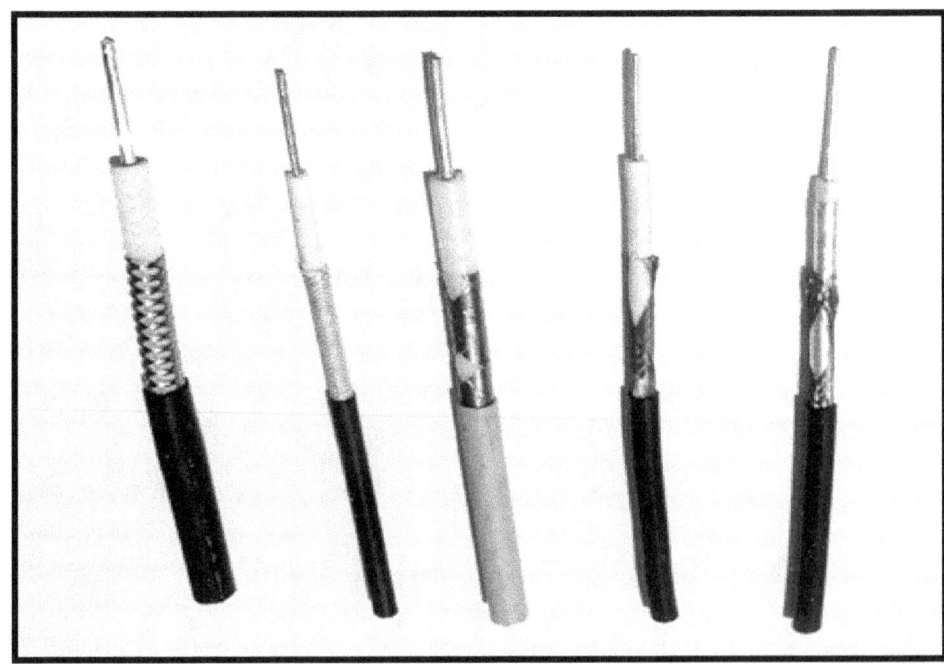

Figure 18: Examples of ported coax cables.

3-20

3.3.3 Sources of Nuisance Alarms

The nuisance alarm rate for ported coax sensors is relatively low in a well-designed and carefully installed system, but several environmental and installation factors can affect the rate. (See Figure 19.) The major source of nuisance alarms is surface water from rain or melting snow, which makes proper drainage essential. Both flowing water and standing water that produces ripples when the wind blows will contribute to movement that is targeted by the sensor.

Movement of metallic objects (e.g., automobiles and fences) or dielectric objects (e.g., animals, people, and plants) in the vicinity of the sensor cables can result in alarms. Small animals weighing less than 4 kilograms (9 pounds) generally do not cause alarms, unless they cross the sensor cables in groups. The magnitude of the nuisance alarm rate also depends on the type of sensor, the installation method, maintenance practices, and the sensitivity setting.

Figure 19: Sources of nuisance alarms for ported coaxial cable detection systems include nearby moving vehicles, larger animals, large and wet plants moving in the wind, and moving water, as well as the ripples caused by wind blowing across standing water.

3.3.4 Characteristics and Applications

The following list describes the characteristics and applications of ported coax sensors:

- They have volumetric, covert detection capabilities. The intruder must guess where the detection envelope is in order to defeat the sensor by bypassing it.

- They are often used where esthetic considerations dictate a sensor technology that is unseen by the general passerby.

- They are very effective in detecting the crawling intruder and therefore are often used as a complementary sensor to other sensor technologies, such as microwave, e-field, or active infrared technologies, that may have a weakness to this method of attack.

- The pulsed ranging systems provide the added benefit of allowing the use of a lower number of processors to monitor multiple perimeter sectors.

- Even though the detection envelope of the ported coax sensor can be relatively wide in volume (depending on cable spacing), it is limited in height, and any method allowing an intruder to bridge over it could result in a successful defeat of the system. If not protected, nearby fences and utility poles can be used for this purpose. For this reason, it is not generally recommended to install this sensor as a stand-alone system around a perimeter for high-security applications.

- The installation of this system requires extensive trenching. In addition, once the system is installed, any needed maintenance of the cables or decouplers buried in the ground could require significant effort to locate and repair these components.

- Even though the system is terrain-following and does not require a straight line of sight, site preparation is very important to ensure the provision of proper drainage to reduce nuisance alarms caused by runoff.

- Depending on location, installing a good fence system to isolate the cables from animals is also important to reduce the nuisance alarm rate to an acceptable level.

3.3.5 Installation Criteria

Ported coaxial cable systems should be installed according to the manufacturer's specifications. Generally, this type of system can operate in longer segments than other detection systems. However, it is recommended that detection zones be restricted to segments of 100 meters (328 feet) or less to facilitate assessment. The system is terrain-following and can be curved around corners as long as the minimum radius specifications are observed. The maximum and minimum separation of the transmitter and receiver cables can vary from approximately 2 meters (6.6 feet) to being co-located in a single trench. The separation is dictated by the clear zone available and proximity to objects that may interfere with the sensor's detection envelope. The lines are generally buried at a depth of .23 meters (9 inches). (See Figure 20.)

The installation of ported coaxial cable perpendicular to buried metal conduit for electrical cables or metal pipes used for water or storm drains may degrade detection capabilities or cause nuisance alarms. Separation distances dictated by the manufacturer should be strictly observed. Soil conductivity should be considered when installing this type of sensor as it may reduce the detection volume. Soil found to have relatively high conductivity may cause the detection field to be reduced. Highly conductive soil contains concentrations of iron or salt. Moving objects in the zone of detection such as foliage, flowing or standing water, and grasses may create nuisance alarms. Rodents can chew through ported coaxial cable.

Figure 20: Typical ported coax cable system installation.

Sensor locations should be carefully selected to prevent nuisance alarms from such sources as personnel and vehicular traffic. Similarly, the cleared area above the sensor should be controlled to prevent the placement of objects within the area, even temporarily, which would degrade the detection zone. The ported coax cables should be installed on well-drained terrain cleared of trees, tall grass, and bushes. Freezing or thawing of the ground may affect system sensitivity. Neither the transmitter nor the receiver lines should be mounted above ground.

The system should be installed, relative to perimeter fencing, so that the transmitter and receiver lines are positioned to prevent someone from avoiding detection by jumping over the electromagnetic field. Typically, the distance between chain-link security fencing with an overall height of 2.4 meters (8 feet) and the center of the detection zone should be a minimum of 2.4 meters (8 feet).

Fixed metal objects and standing water distort the radiated field, possibly to the extent of creating insensitive areas with no, or very low, detection. Nearby metal objects and utilities should be located outside the detection volume. This includes aboveground fences and poles, as well as underground water and sewer lines and electrical cables installed close to the surface. (Refer to Figure 21.)

Manufacturer's instructions should be followed when installing cable across concrete or asphalt areas. Particular attention should be paid to the binding agent and the application of epoxy over the cable groove after the cable is installed in the concrete or asphalt.

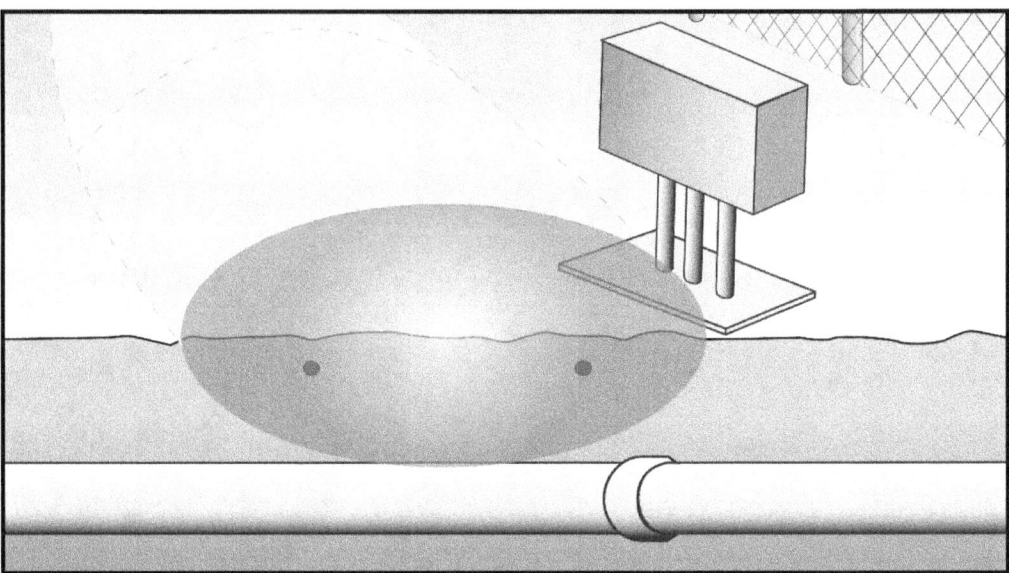

Figure 21: Metal objects or utilities located within the detection volume of a ported coax system may create areas of poor detection.

Criteria for ported coaxial cable system installation include the following:

- Ensure that the ground in which a ported coaxial cable system is buried is firm and is not subject to movement.

- Ensure that drainage is adequate because surface water (standing or flowing) can cause ported coaxial cable systems to generate false alarms.

- Rodents can chew through ported coaxial cable—be aware of their burrows.

- Avoid intersecting irrigation pipes and power lines with the coaxial cable, or ensure that separation distances specified by the manufacturer are observed.

- Note that the detection zone may be elongated at curves.

- In rare cases of extreme soil conditions (either very sandy or highly conductive), soil conductivity tests may be warranted to ensure that performance of the system is not compromised.

3.3.6 Testing

A regular program of testing sensors is imperative for maintaining them in optimal operating order. Three types of testing need to be performed during the life of a sensor: acceptance testing, performance testing, and operability testing.

3.3.6.1 Acceptance Testing

When a ported coaxial sensor is first installed, it should be tested in order to formally "accept" the sensor as part of the physical protection system. Acceptance testing consists of two parts:

(1)   A **physical inspection** to ensure that the sensor was installed properly consists of the following:

- Verify that the installation matches the installation drawings, which should follow the guidance provided by the manufacturer.
- Verify sector intersection spacing.
- Verify that signal and power wires are routed in the conduit.
- Verify proper power levels (voltage and amperage).
- Verify correct wire connections.

(2)   A **performance test** should establish and document the level of performance. Refer to the performance testing procedure (below) for a description of the recommended tests.

3.3.6.2   Performance Testing

A ported coaxial cable perimeter detection system should be capable of detecting an individual passing over the transmitter and receiver wires, whether the individual is walking, running, jumping, crawling, or rolling.

The probability of detection is affected by the installation configuration of the cables, the system processor settings, soil conductivity, intruder orientation and speed, and the proximity of metallic objects.

Testing should ensure that the device is capable of detecting the defined target with a site-specific probability of detection of 90 percent and a confidence level of 95 percent. The system should detect a person crawling, running, or jumping through any area of the detection volume. As such, performance testing should include these defeat methodologies. Testing has shown that a person using a "duck walk" profile (walking in a crouched position) to pass through the detection volume produces the smallest disturbance to the system. (Refer to Figure 22.) Sensor parameters should be set to detect this type of intrusion.

Figure 22: The "duck walk" seems to produce the smallest disturbance in a ported coaxial system and is therefore ideal to use to test the system.

3.3.6.3 Operability Testing

Operability tests should consist of simple walk tests. The testing individual walks through the expected detection zone of a sensor and confirms that the alarm has been received at the alarm station. Operability tests should include a search for any evidence of damage to the sensor or tampering with the device.

3.3.7 Maintenance Requirements

Because the ported coaxial cables are buried and largely inaccessible after installation, maintenance to the system is minimal and involves the processor and the environment where the system is installed. Periodic checks should be conducted at least every 6 months and include the following:

- If applicable, check terminations and/or decouplers between sectors of sensor cable that are located above ground for signs of wear or leakage.

- Inspect the processor enclosure for any physical damage, water damage, corrosion, and ingress of insects.

- Inspect the cable connections (coax, signal, or power) to ensure that they are tight.

- Check the processor electrical ground (earth ground) connection for continuity and corrosion.

- Check the input power for proper voltage. Check the battery status (if applicable).

- Inspect the areas above the sensor cable for vegetation, debris, water accumulation, erosion, and settling of trenches. Correct as necessary.

- Verify processor parameter settings to ensure that no undocumented changes have been made.

3.4 <u>Active Infrared Sensors</u>

3.4.1 Principles of Operation

Active infrared sensors are infrared beam-break sensors that detect the loss or significant reduction of infrared light transmitted to a receiver. Infrared light is invisible to the naked eye (though some digital or video cameras can detect it.) In its simplest form, an active infrared sensor consists of a single infrared transmitter that illuminates a single infrared receiver. In most security applications, an active infrared sensor will be composed of two columns of multiple infrared transmitters and multiple infrared receivers. This arrangement can provide a detection volume of significant height. The actual volume of detection is defined by all the beams (cylinders of infrared light) between the transmitters and the receivers and has the diameter of the receiver and transmitter's optical lens. In a multibeam active infrared sensor, the actual volume of detection is composed of the numerous infrared beams that shine between the two columns. (See Figure 23.) The columns can be separated by up to 152 meters (500 feet) or more for some available types. The lowest beam is generally located within

6 inches of the ground to detect crawlers; the spacing of the upper beams can generally be further apart.

Figure 23: Multibeam active infrared sensor columns. (Red lines represent beams of infrared light.)

Detection is based on blocking one or more of the beams for a specific period of time. Shorter measured beam interruption times are discounted to reduce nuisance alarms from small birds or blowing debris. Some models allow for user adjustment of this beam interruption time while others automatically vary the time depending on the number of beams interrupted.

The infrared light is generally modulated and sequenced to reduce sensitivity to other light sources, which allows the sensor to operate in daylight as well as night. The sequencing for some models allows the use of multiple sensor sets without mutual interference, in case they happen to be within the field of view of an adjacent sensor set.

3.4.2 Types of Active Infrared Sensors

Active infrared sensors vary in size, beam configuration, range of operation, and operational features. They typically operate in ranges that allow them to be compatible with the sector lengths used in exterior perimeter applications. A single-beam unit will have limited application because of the small, single-beam detection volume. Many manufacturers provide sensors that can be combined to provide whatever detection height is required for an application.

Some models can communicate to a remote laptop or other computer, a capability that allows them to be remotely configured or adjusted. Some models also have the capability for the user to turn off the lowest beam (for example, if the ground is covered by snow) to allow the remainder of the beams to operate normally. Most models have sealed beam assemblies or heaters that keep frost and condensation from forming on the optical covers and lenses. Some units can be sequenced so they will operate if their transmitters are in the field of view of an adjacent sector's receivers. These and other features of the different models available should be considered, depending on the specific application and operational environment.

Figure 24 below provides examples of the available designs of active infrared systems with varying transmitter, receiver and beam configurations.

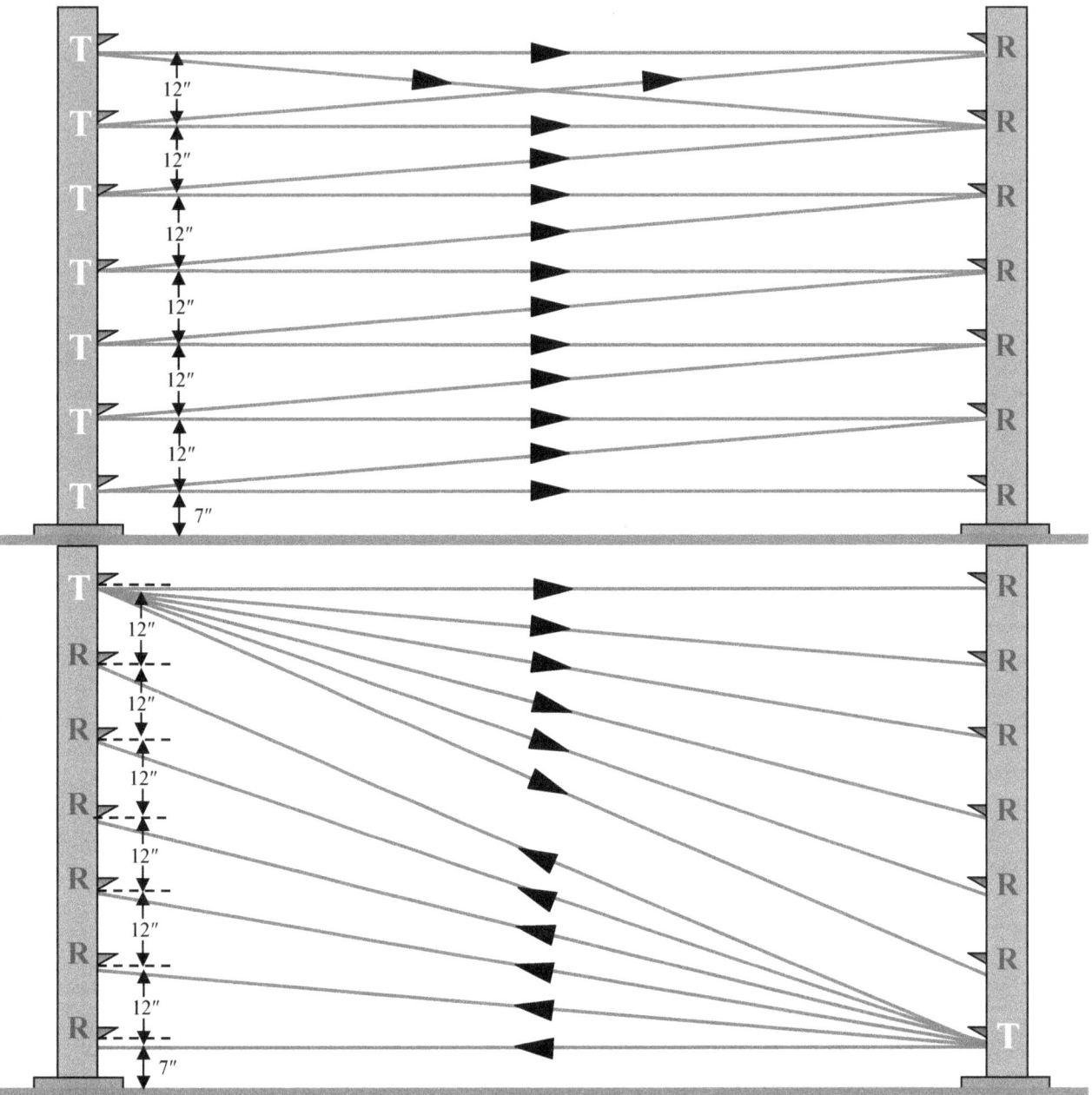

Figure 24: Active Infrared sensors vary in beam configuration. (T = Transmitter, R = Receiver)

3.4.3 Sources of Nuisance Alarms

Nuisance alarms can be caused by anything that can block or significantly obscure the infrared beam. Nuisance alarms from animals large enough to block the bottom beam can be minimized by installing the sensor between two fences. A single bird flying through an infrared beam will generally not cause an alarm, but a flock of birds may cause sufficient beam blockage to cause nuisance alarms.

Blowing debris that blocks the bottom beam can cause alarms. Snow accumulation to a depth that covers the lower beam will cause a constant alarm that must be cleared by removing the snow or disabling the lower beam (on sensors with that capability). A disabled lower beam could allow an intruder to crawl under the sensor undetected; once the snow has been removed, the sensor should be restored to full operation.

Heavy snow or fog thick enough to obscure visibility from transmitter to receiver may cause nuisance alarms. Bright lights (such as dawn or dusk sunlight during certain times of the year) that shine directly into the receiver can cause a nuisance alarm or a failure alarm. Some models have modules that contain both a transmitter and a receiver. In these models, the beam actually consists of two beams, one going in either direction. Both beams must be blocked or interfered with to cause an alarm.

3.4.4 Characteristics and Applications

The characteristics of an active infrared sensor include the following:

- Line-of-sight operation

- Volumetric detection that resembles line detection

- Beam-break sensors (not field disturbance sensors)

- Impervious to motion around it

These characteristics make active infrared sensors especially useful in confined areas or in areas where there is much activity nearby. Thus, active infrared sensors are successfully used in entry control portals to substitute for a sensor that would block traffic flow or would have a high nuisance alarm rate because of activity in the area. They can operate over any surface and can "see" through chain-link fences, provided that a fence post or pole does not block any beams.

Active infrared sensors are successfully used in perimeter applications. Because of their uniformity and detection height, active infrared sensors make a good complementary sensor to other types of sensors that have good crawl detection but lack the significant detection height that can be achieved with an active infrared sensor.

Most models can operate well in a typical PIDAS sector length of 100 meters (328 feet); however, if local weather conditions have the potential to produce periods of reduced visibility, the installed operating distance should be reduced to 80 meters (approx. 264.2 feet) or less. (This is a good design practice for the alarm assessment (video) system as well, because video is also impacted by poor visibility.) Those models with a shorter operation range can employ multiple sets of active infrared sensors combined to accommodate the required sector length.

Some active infrared sensors can be applied to special applications. They can be adapted to provide detection across doorways or windows, or across critical access doors. They can be installed across pedestrian and vehicle access openings in a PIDAS. Because of their modular nature and narrow detection width, active infrared sensors can be adapted to many areas that would be difficult or impossible for other types of sensors to monitor well.

Because the detection characteristic of the active infrared sensor is a beam-break rather than a field-disturbance sensor, its detection volume can be more readily identified by a potential adversary. It can be closely approached without concern of detection as long as the beams are avoided. This means that if the detection height is configured to be 1.8 meters (6 feet) or less, an intruder aided by a step ladder could easily defeat the sensor by climbing the ladder and jumping over the top beam. Luckily, active infrared sensors can be configured to be 3.7 meters (12 feet) or more high. Jumping from this height becomes much more difficult; therefore, if an active infrared sensor is fairly short in height, it should be combined with a field disturbance sensor that can prevent a close undetected approach by an intruder, as well as detect attempts to crawl under the active infrared's lowest beam. The height of the zone of detection of an active infrared system should account for the design and configuration of physical barriers within the vicinity of the zone of detection to ensure that the system provides protection from bridging.

Because active infrared sensors are line-of-sight sensors, the ground between the transmitter and receiver columns must be relatively flat and planar (lying in a single geometric plane). Any low areas can be used to crawl undetected under the bottom beam. Even if the ground is flat and level, if a trench can be dug without detection, it may be possible for an intruder to quickly create a path onto the site under the bottom beam. It is recommended that if an active infrared sensor is used alone, a hard surface, such as a concrete sill or sidewalk, should be installed under the beams to prevent trenching, thus minimizing the potential for the system to be circumvented.

Use of an active infrared sensor in locations that experience frequent deep snow, drifting snow, or dense fog must be carefully considered. During times that the lower beam is covered and disabled, crawl defeat is possible unless the active infrared sensor is used in combination with another sensor that can detect through the snow. In addition, during periods of dense fog, the sensor may go into constant alarm. During these periods, a complementary sensor that is not normally affected by fog would be very important in the detection of intruders.

3.4.5 Installation Criteria

A cleared area must be created around the sensor system location. Ideally, this area should be protected from passing pedestrians or animals. Isolating the detection equipment between fences provides this protection, as well as other security benefits. It is critical that weeds be eliminated in this area to avoid attracting animals and to prevent tall plant growth from blocking the infrared beams.

The active infrared sensor should be mounted on a sturdy foundation that will not heave as the ground freezes or move in the wind. Most models have a field of view that will tolerate very small motion, but movement significant enough to misalign the beams will increase the sensor's susceptibility to nuisance alarms.

All alarm, communication, and power wiring to the sensor should be in buried conduit to reduce the risk of accidental damage or malevolent tampering.

Active infrared sensors should be installed over ground that is flat and even. In a stand-alone configuration, the sensor should have a detection height of 3.7 meters (12 feet) or more and should be installed over a hard surface made of concrete or other material to prevent the digging of a trench beneath the lower beam. In a stand-alone installation, the lower beam should be no higher than 6 inches, depending on the size of the optics. Therefore, a crawling intruder must interrupt the beam sufficiently (typically 98-percent beam block) to cause an alarm at any location between the transmitter and receiver. Some models will not have a beam this low. The alternatives are to use an active infrared sensor in a complementary combination with another sensor that will detect a crawler or to install a hard surface beneath the sensor that is high enough to reduce the distance to the lower beam and, hence, allow for crawler detection.

In locations that are susceptible to freezing, a sensor heater option is available for many models. The heater is typically controlled thermostatically and requires considerably more power than the sensor electronics. Care should be taken to size the wire appropriately for the electrical current requirements of the sensor heater.

To be effective, all perimeter sensors should be installed in continuous lines of detection. For the active infrared sensor, sensor overlap should be provided. The layout must be designed in such a manner to prevent gaps large enough for an intruder to maneuver around the beams of adjacent sectors. One method of intersecting sectors is to "basket weave" the active infrared sensors. An intersection can also be made by overlapping the beams in parallel, such that the beams of each sector pass very close to the transmitter or receiver column of the adjacent sector. A transmitter should face the adjacent transmitter and the receiver should face the adjacent receiver to avoid interference. Figure 25 illustrates these two methods.

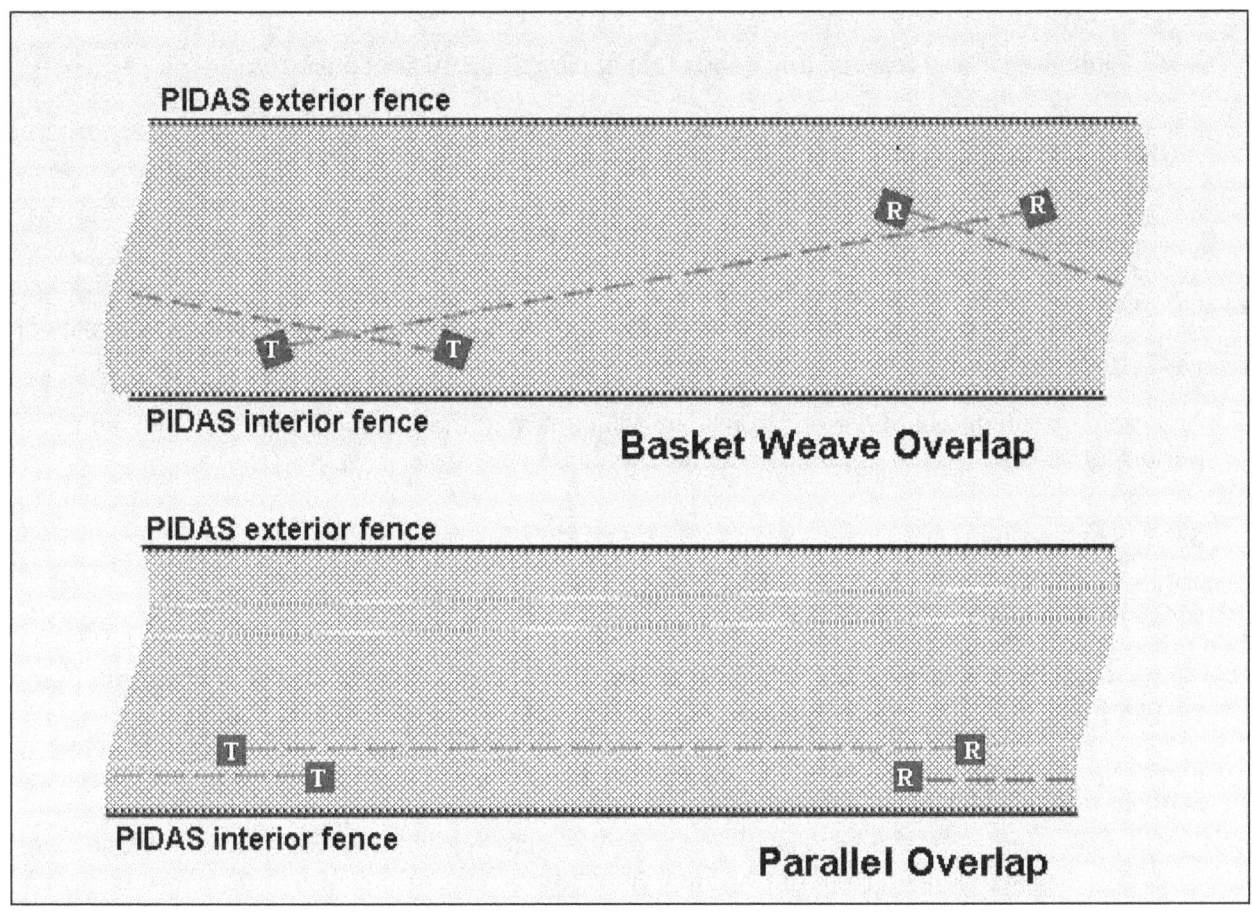

Figure 25: Active infrared sector intersection methods.

Another method, a variant of the parallel method, is to mount the transmitter and receiver columns from adjacent sectors next to each other, leaving no gap large enough to allow an intruder to pass between them. Once installed, the active infrared sensor modules must be aligned to their matching receiver or transmitter in the opposite column. Poor alignment can result in a poorly received infrared signal level. This, in turn, can result in increased nuisance alarms.

3.4.6 Maintenance

Active infrared sensors should be maintained in accordance with the manufacturer's recommendations. Routine maintenance should involve a physical inspection of the sensor and the clear area where the sensor is mounted. Plant growth, debris, or anything that is unusual should be removed. Maintenance includes periodic cleaning of the sensor's optics. The physical inspections can be combined with the scheduled testing of the sensor. However, an increase in nuisance alarms indicates the need for physical inspections; increased nuisance alarms should be investigated as soon as noted.

Routine inspections should include the following:

- Inspect for any weed growth and sterilize the soil or remove any growth near the sensor.

- Inspect for and remove any debris that could cause nuisance alarms.

- Inspect the lens or lens covers for any dirt or debris, including insects or spider webs; clean as needed.

- Inspect all mechanical parts for any sign of damage or tampering.

- Look for any areas of erosion or damage to the hard surface under the lower beam (if used) that could eventually increase the gap beneath the active infrared sensor's lower beam.

Any major replacement of the electronics or adjustment of the sensor should be followed by a performance test. All maintenance actions should be recorded.

3.4.7 Testing

3.4.7.1 Acceptance Testing

When a sensor is first installed, it should be tested in order to formally "accept" the sensor as part of the physical protection system. Acceptance testing consists of two parts:

(1) a **physical inspection** to ensure that the sensor was installed properly, as follows:

- Verify that the installation matches the installation drawings, which should follow the guidance provided by the manufacturer.
- Verify sector intersection spacing.
- Verify that signal and power wires are routed in conduit.
- Verify proper power levels, both voltage and amperage.
- Verify correct wire connections.

- Perform an optical alignment on all units to verify that all modules are operational and oriented correctly. The method of optical alignment is specific for each manufacturer's model.

(2) a **performance test** to establish and document the level of performance.

3.4.7.2 Performance Testing

A performance test should be designed to verify the level of performance of each active infrared sensor through its range of intended function and should include the use of all defeat methodologies applicable to the system. The replacement of the electronics module, a change to the optical alignment, or any adjustment that could affect the unit's sensitivity are examples of system changes that require performance testing. The performance test should include the use of the documented levels of performance from the original acceptance testing to verify that the sensor is still performing adequately.

The following describes a recommended performance test for an active infrared sensor:

(1) A performance test should include a visual inspection of the sensor and of the general area where the sensor is installed. Refer to section 3.4.7 "Maintenance" of this document to perform the recommended routine maintenance inspection.

(2) Record all relevant processor readings. Compare to the last recorded readings, either from the acceptance test or the previous performance test. Each manufacturer's model will have different sensor settings. Make a special note regarding any significant settings that may influence the detection ability or the nuisance alarm rate of the sensor.

(3) Active infrared intrusion detection sensors can have a probability of detection for a walking intruder that appears to be perfect ($P_D = 1$). This is because the sensor is a beam-break sensor and has no direct adjustment for sensitivity. A breaking of the infrared beam for the prescribed time will produce an alarm if the electronics are not defective. It is necessary, then, to confirm that the sensor does produce an alarm output that is communicated to the alarm system in response to a beam break of each beam. Cover each beam and verify that an alarm is produced. Do this at both columns.

(4) Because most units employ a timing feature, it is necessary to confirm that a fast-moving intruder will be detected when penetration occurs. A person should penetrate the sensor moving as fast as possible, while avoiding the bottom beam, which sometimes has a faster beam-interruption time than the other beams.

(5) The area where the sensors overlap should be tested to verify that each attempt to bypass the sensor under test produces an alarm on either the sector being tested or the adjacent sector.

(6) The active infrared sensor can be defeated by bypassing the beams, either by crawling under or by aided climbing and jumping. Pay particular attention to any low places. If used in a stand-alone configuration, verify that the hard surface underneath the lowest beam is intact. Verify that attempts to crawl under the sensor are always detected at the location where the greatest gap exists under the lower beam. If this location cannot be readily identified, distribute the crawl test along the length of the sector being tested.

(7)   If the sensor is used as part of a complementary sensor system and the other sensor is intended to detect the crawler or to detect climbing approaches that use ladders or other aids, then that additional sensor should also be tested to verify that it functions as intended (by detecting what its complementary sensor does not). (This objective should also be part of the other sensor's performance test plan.)

(8)   Optical alignment of an active infrared sensor is a common adjustment required of all models. A misaligned sensor will have an attribute of an increased number of nuisance alarms. If nuisance alarms are excessive, inspect for evidence of animals or debris present, and check the alignment signal level if available on the specific model. Also inspect the optics for dirt or damage. If the alarms occur at certain times of the day and dew or frost might be the cause, check that the heater is operational.

If no cause for the increased nuisance alarms can be identified, perform an optical alignment. After alignment, slowly obscure each beam and identify the one that requires less blockage than the others before alarming. This will indicate that either the transmitter or the receiver module is defective, especially if the beam has already been aligned.

3.4.7.3 Operability Testing

Operability tests are simply "walk tests." The testing individual walks through the detection zone of the active infrared sensor and verifies that the alarm arrived at the alarm display center. It is recommended that this testing be incorporated with an inspection of the perimeter security system. This inspection should note the status of the perimeter fence and other physical barriers and all physical security equipment nearby. The inspection should look for evidence that any component has been tampered with or damaged. Appropriate actions for repair should be taken if needed, and indications of tampering should be reported, investigated, and documented.

3.5   Taut Wire Sensors

3.5.1 Principles of Operation

An installation of taut wire sensors consists of multiple strands of twisted wires strung between dedicated posts, known as a freestanding installation (see Figure 26), or attached to an existing fence (see Figure 27). The twisted wires are generally made of a high-tensile strength barbed wire (**not** a farm-grade barbed wire). These wires function as a spring over a wide temperature range. A switch or sensor is attached to each wire and located in the middle of the wire spanned between two anchor posts. The distance between anchor posts can be up to 100 meters (328 feet), depending on the type of sensor and the sensitivity required. The number of wires and their spacing are critical to the effectiveness of this sensor.

Figure 26: A freestanding taut wire installation for environmental testing.

Taut wire sensors will register a detection and alarm when one or more sensors sense a deflection of the wire(s) in a horizontal direction. This horizontal deflection is caused either by a certain amount of pressure applied somewhere along a wire's length (see Figure 28) or by a wire being cut or broken.

Within the categories of intrusion detection sensors, a taut wire system is a terrain-following, point-of-detection and visible sensor.

Figure 27: A taut wire sensor system installed on an existing fence. Note the application of taut wire to the outriggers in this test application.

Figure 28 A tester pushes down with moderate pressure on one of the wires of a taut wire installation. Note the deflection of the switch (sensor).

3.5.2 Types of Taut Wire Systems

The major difference between the various taut wire systems is the type of sensor/switch that is used to detect a deflection in a wire. The two most common switches on the market today are mechanical and electromechanical. The mechanical switch is a simple device whose threshold of detection can be adjusted only through the adjustment of tension on the individual wires. The electromechanical switch can control the threshold of detection through the use of software to adjust the detection threshold for each wire's tension, which allows for easier adjustments for greater ease of implementation and use. Types of taut wire electromechanical switches available on the market include strain gauge, piezoelectric, fiber optics, and resistive rubber.

3.5.3 Sources of Nuisance Alarms

The primary advantage of a taut wire sensor system is that, if correctly installed and maintained, there are few sources of nuisance alarms. This statement should **not** be interpreted to mean that taut wire is an ideal sensor for most applications. (Refer to the "Characteristics and Applications" Section below.)

One possible source of nuisance alarms, though fairly rare, is a serious ice storm. If large amounts of ice build up on the wires, physical damage to the system may occur.

In some areas of the country, the facility experiences a wide variety of temperature fluctuations (from -20 degrees Fahrenheit (F) to 100 degrees F) over the course of a year. This large temperature fluctuation can change the tension of the wire, causing sensitivity concerns and may require additional maintenance because of the stretching and shrinking of the wire. A facility in this type of location may need to perform readjustments or retensioning of the wires on a seasonal basis to provide a uniform probability of detection.

A highly corrosive environment, such as near the ocean, can cause increased maintenance and reliability concerns if the taut wire system is not purchased in stainless steel or aluminum versions. Also, during high winds, a large piece of heavy debris (such as heavy cardboard or a sheet of plywood) flying against the wires could cause a nuisance alarm. Refer to Figure 29.

3.5.4 Characteristics and Applications

The following list describes the characteristics and applications of taut wire fence sensors:

- Very low nuisance alarm rate; small animals (rabbits, rodents) and weed vegetation do not typically cause alarms

- Terrain-following (to a degree)

- Not affected by most weather conditions (except for heavy ice buildup)

- Works effectively during fog conditions

- Installed on the outside of the inner perimeter fence as a validation of an alarm condition in conjunction with a different sensor type

- Can be installed along outriggers of a fence

- May be installed along the edge of roofs as both a barrier and a sensor

- Can be applied in highly corrosive environments if all components are manufactured with stainless steel and aluminum

- Can be mounted on an existing chain-link fence with minor modifications

- Can be combined with a fence disturbance sensor as an outrigger system to detect climbing intrusions, while relying on the fence disturbance sensor to detect cutting intrusion attempts (less expensive than a full taut wire sensor installation)

- Optimal performance when the taut wire sensor serves as part of a multisensor PIDAS

Figure 29: Examples of sources of nuisance alarms for taut wire fencing include ice forming on the taut wires, flying heavy debris, or a location that experiences weather extremes—very cold winters and very hot summers.

- Depending on the sensor and the intermediate and anchor post configuration, a taut wire system may be able to be bridged using a ladder without causing a detection.

- Clamping and cutting the wires or slowly separating the wire to create a passable opening could allow an intruder access without detection (although this is not an easy thing to do).

- If noncorrosive components are not used, wires and components could corrode in a highly corrosive environment, which could affect the friction between wires and their mountings, thereby decreasing the sensitivity of the system.

- Tunneling/digging under the taut wire may be a concern if an adequate concrete curb or a sensor designed to detect a crawling intruder, such as a ported coax sensor, is not installed beneath the lowest wire.

- If anchor and/or sensor posts are poorly installed, slight movements of the posts can cause a significant nuisance.

- A taut wire system should not serve as the single sensor in a PIDAS.

- A taut wire sensor can be installed on gates, but it is challenging. The use of another technology, such as an active infrared sensor, on a gate is much preferred.

- As with most sensors, a taut wire system should not be installed on the outside of the outermost PIDAS fence, as the system could be easily bridged.

- A taut wire system should not be installed on the inside of either PIDAS fence because an adversary could climb the outside of the fence, then jump down without the taut wire system alarming.

3.5.5 Installation

Manufacturer's specifications should be followed for installation.

Installation of a taut wire system may be freestanding (where the wires are mounted on their own dedicated posts), or the system may be attached to an existing sturdy fence. For a freestanding system, it is critical that the outer-most **anchor posts** for each taut wire system be installed with sufficient bracing to withstand the tension created by the wires. The bracing should not provide an advantage to an adversary attempting to climb the post to bypass the IDS.

The anchor post will generally be a sturdier post than the sensor and intermediate posts in the system.

The wires of the taut wire system are securely attached to each anchor post in various ways depending on the manufacturer. For short sectors, springs are usually incorporated at the anchor post to provide enough movement at the sensor to enable an alarm.

A **sensor post** is installed at the midpoint of the sector length (which is usually 50 or 100 meters (164 or 328 feet), depending on the system manufacturer). Sensor/switches will be mounted along this pole, and the wires are attached to these switches.

Additional posts, referred to as **intermediate posts,** are installed about every 3 meters (approx. 9.8 feet) between each anchor post and the sensor post. A slider coil the length of the post is installed at these locations to provide a support to, and to maintain the spacing for, the sense wires. The installer slides a **fixing bar** down between the post and the slider coil such that the wires remain in position but are still free-moving.

If barbed wire is used, the installer should be certain that a barb from the barbed wire does not catch on any of the coils; the wire must remain "free" between the anchor post and the sensor post. One or more individual barbs from the barbed wire may need to be removed to prevent excessive friction.

Depending on the threat facing the facility, concrete curbing may be installed beneath the bottom-most wire to discourage digging underneath the taut wire fence.

Typical installation guidelines require approximately 100 pounds of tension on each wire. A typical taut wire system will alarm when 25 pounds or more of pressure is exerted on at least one wire. This pressure can result from a person pulling on a wire in any direction, from a wire being climbed on, or from a wire being cut, which releases the tension on one side of the sensor.

A taut wire system is terrain-following to only a certain degree. Over the length of a taut wire unit, there can be no more than a total of 15 degrees of change in the elevation of the wire between all posts, with the wire always remaining parallel to the ground. That 15 degrees includes change both up and down. If there is more change in the angles than that, the friction between the wires and the coils holding them becomes too great, such that the wires are not adequately free-moving to detect an alarm condition.

3.5.6 Testing

3.5.6.1 Acceptance Testing

Acceptance testing is the process that a site must go through after an installer has completed the installation of the sensor system but before the system is accepted and used operationally.

Acceptance testing (for IDSs) consists of two parts:

(1) A complete and positive performance test (see below).

(2) Before acceptance testing, the newly installed system should be thoroughly examined to ensure that it has been installed according to manufacturer's specifications and the detailed engineering drawings for the site-specific design/installation of the taut wire system. A facilities engineer will need to ascertain that commonly accepted practices have been followed, such as: Was the system wired correctly? Are poles installed at 90 degrees? Are wires parallel with the ground and each other? Are voltages correct?

3.5.6.2 Performance Testing

Performance tests for this system should be performed on the wires on each side of a sensor because the lowest sensitivity of the system will be located at the furthest point away from the sensor post. Testing should be concentrated at or near the anchor post. Other tests can and should be performed at other locations along the wires to verify uniform sensitivity. However, testing of the wires should not be done close (approximately within 10 feet) to the sensor itself as damage can occur to the sensor.

Two types of tests can be performed on the sensor without damaging the system. Each of these methods tests that the deflection of the wires from a person either climbing the wires or placing a ladder against the wires will cause an alarm. No actual cutting of the sense wires should be performed.

The ladder test consists of placing a ladder against the wires and an individual climbing the ladder to a point where sensor activation occurs. The alarm should generally occur before the knees of the tester are near the top of the fence but the test should be terminated as soon as an alarm is received to prevent damage to the sensors. It is a good idea to provide a local alarm annunciation during testing to facilitate this.

To test the individual wire deflection, a yard stick can be used to measure the distance that each wire can be moved before an alarm occurs. Providing a method to hang the yard stick from above the wire being tested can facilitate this. Refer to Figure 30 for an illustration of this measurement technique. The distance that any single wire should be able to be deflected before receiving an alarm is based on site-specific requirements; however, a deflection of no more than 3 to 4 inches should be tolerated.

Figure 30: A hanging yardstick allows for easy measurement of the deflection possible on a taut wire sensor before an alarm occurs.

For taut wire systems utilizing electromechanical devices for monitoring the wires, any software setting or sensitivity parameters that are adjustable in the processors should be measured or verified and documented. Also, any processor box located at the fence should have a tamper switch that is tested as part of the semi-annual performance test.

Deflection results should be documented to be able to monitor for system changes or degradation over time.

During periods of extreme cold weather, it may take some time for the mechanical sensor switches to return to the normal neutral position after activation. This should be considered when considering multiple tests of the same zone.

3.5.6.3 Operability Testing

The standard operability test for this system is simply a matter of pulling up or down on one of the wires of a particular sector until an alarm is received at the alarm monitoring station.

Because of the number of sensors (i.e., wires) that make up the taut wire system, it is wise to provide some method of ensuring that a different wire is tested each time the test is performed. This ensures that, over time, all of the wires and sensors will be tested for operability.

3.5.7 Maintenance

Maintenance to the taut wire system is essential for consistent performance. During the weekly operability testing of the system, the tester should be observant of any abnormalities that may exist. At a minimum, the inspection should entail the following:

- Remove any foreign objects that may have blown into or otherwise been lodged on the wires.

- Remove any vegetation growing onto the wires.

- Inspect the ground beneath the system for signs of erosion or digging by animals.

- Inspect the condition of the slider coils, and ensure that the fixing bar is in place and is secure.

- Ensure that the wires are not jammed or snagged at the slider coils or sensor post and that no barbs are within 2 to 3 inches of the slider coils.

- Inspect the condition of the sensors to ensure that components are not cracked or damaged and that the wires are securely attached.

- Inspect wires for condition, spacing, and tautness.

- Perform any other inspections or maintenance specified by the manufacturer.

- At least once per year or whenever physical inspection warrants, the tautness of the wires should be verified using a dynamometer to ensure tension is within manufacturer's specifications.

3.6 Fence Disturbance Sensors

3.6.1 Principles of Operation

A fence disturbance sensor is designed to be mounted on a chain-link fence to detect disturbances of the fence, such as the noises or motions generated when an intruder attempts to climb over or cut through the fence fabric. (Refer to Figure 31.)

Figure 31: This diagram illustrates the deflection caused by climbing on chain-link fence fabric; this disturbance of the fabric is one of the ways that fence disturbance sensors work.

A climber generates two types of disturbance while climbing the chain-link fabric: a slow, low-frequency motion of the fabric movement and a higher frequency, impulsive shock or noise caused by pieces of the fence striking or rubbing against each other. Cutting the fence fabric also generates an impulsive shock. Some types of fence disturbance sensors have dual processors so that they can distinguish between the slow, low-frequency motion of the fabric and the high-frequency shock noises.

When someone or something causes a significant disturbance to a fence, a switch will activate or a signal will cross a threshold, which is considered to be an event.

Most fence disturbance sensors require a specified minimum number of events to occur within a specified time in order to generate an alarm condition. For example, a facility could specify that an event must be detected five times within half a minute before an alarm is reported. The decision for these parameters should be driven by the threat, physical protection program goals and the physical characteristics of the fence (i.e., height and wire mesh size). A lower number of required events would increase the probability of detection but increase the nuisance alarm rate as well. Requiring a higher number of events to generate an alarm would cause the opposite (i.e., a lower probability of detection but improved (lower) nuisance alarm rate).

The actual number of events over a specific period of time required to cause an alarm at a particular facility is generally considered sensitive information since knowledge of these parameters would give an adversary the advantage needed to possibly defeat that system.

3.6.2 Types of Fence Disturbance Sensors

3.6.2.1 Mechanical Fence Sensors

Mechanical fence sensors operate on the principle that fence movement will cause a switch to open or will close a set of contacts. One of two approaches is generally used. First, mercury switches are used to detect side-to-side movement or tilting of the fence and signal processors employ count and time criteria to differentiate between intrusions and nuisance indications. The second method employs a mass on a set of contacts. The system detects impulsive fence movements of sufficient magnitude to cause the mass to move off the contacts momentarily. Generally, this method also uses count and time criteria to generate an alarm indication. Such sensors are positioned on the fence at regular intervals.

3.6.2.2 Electromechanical Fence Sensors

Electromechanical fence sensors employ individual point transducers to detect fence motion. The point transducers produce an analog signal instead of a switch closure and use an electronic signal processor to extract alarm information from the signal. Like mechanical fence sensors, these sensors are positioned on the fence at regular intervals. For high-security applications, the transducers are placed on every pole or on the fence fabric itself between each pair of poles.

Two types of electromechanical transducers are used: geophone and piezoelectric. The principles of operation for each type are given below.

3.6.2.2.1 Geophones

A geophone transducer uses the principle of a conducting loop moving in a fixed magnetic field to generate a voltage. Either the coil or the magnets are mounted on a spring structure so that, when the body of the geophone moves, there is relative motion between the coils and the magnets, causing a voltage output.

Geophones are sensitive to movement along one axis only, and thus they are mounted on the fence so that they are sensitive to the vertical movements of the fence. This orientation helps to eliminate nuisance alarms caused by wind-induced fence movements.

3.6.2.2.2 Piezoelectric Sensors

Piezoelectric transducers detect fence motion by employing crystals that exhibit the piezoelectric effect: when the crystal is deformed, a voltage appears across the crystal. The crystals are mounted in a housing in such a way that fence motions are detected. The transducers are sensitive along a major axis and are generally mounted to be most sensitive to vertical fence motion.

3.6.2.3 Strain-Sensitive Cable

Strain-sensitive cable is a transducer that is uniformly sensitive along its entire length. It is specially designed to produce an output voltage when the cable is moved. An electric cable uses a polarized dielectric to become microphonic. Another cable type uses special construction to enhance voltage output caused by the triboelectric effect. The triboelectric effect is an electrical phenomenon in which certain materials become electrically charged by friction after being rubbed together. A third type resembles a distributed geophone with magnetic material surrounding sense wires in one cable. Each cable type is fastened directly to the fence using wire ties so movement of the fence fabric is coupled directly to the transducer cable. Figure 32 shows a typical installation of strain-sensitive cable.

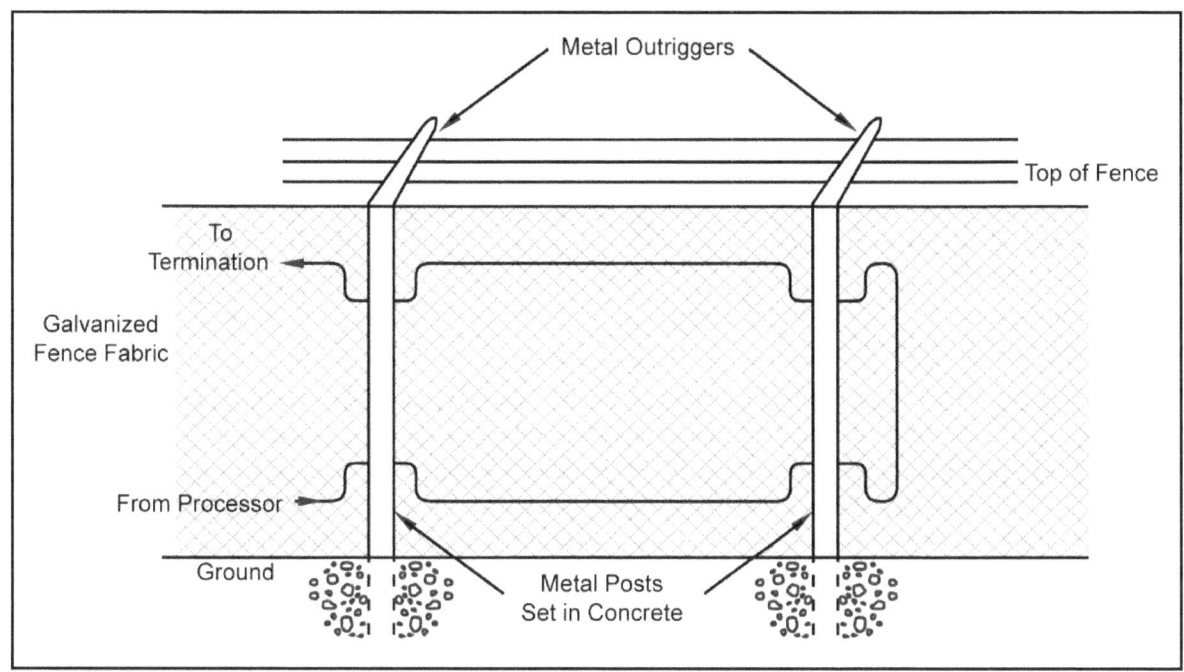

Figure 32: Typical installation of strain-sensitive cable.

3.6.2.4   Fiber-Optic Cable

Fiber optics is the class of optical technology that uses strands of optically pure glass as thin as a human hair to carry digital information over long distances. A light source, such as a light-emitting diode (LED) or laser diode, is coupled to one end of the fiber, while a receiver (such as a photo transistor or similar device) is coupled to the other end of the fiber. The two major categories of fiber-optic sensors are continuity sensors and microbending sensors. Major advantages of fiber-optic cable include immunity to radio and electromagnetic interference and durability in conditions of changing temperature and humidity.

Since the fiber does not have to be straight to reflect the light from one end to the other, it can be installed in various configurations on a fence or be buried in the ground to sense a disturbance. The light diffraction (speckle) pattern and the light intensity at the end of the fiber are a function of the shape of the fiber over its entire length. Motion, vibration, or pressure of the fiber induces modal differences causing a phase shift in the light. Sophisticated processors can detect these phase shifts or minute changes in the light patterns being sent down the cable. These changes can be characterized by the processor as those being produced by someone cutting or climbing the fence, and when the right criteria are met (as set in the parameters of the processor), an alarm will be generated.

Some processors have the capability of allowing long lengths (up to tens of kilometers) of nonsensing cable (single-mode fiber) to be placed between it and a length of sensing cable (multimode fiber). This allows the processor(s) to be located in a central location a long distance from the fence being protected.

3.6.3 Sources of Nuisance Alarms

The tension of the fence fabric and the general condition of the fence can affect both the performance and the number of nuisance alarms that will be experienced. Tight fabric enhances the transmission of noises along the fence, making the slight noises caused by a climbing or cutting intruder easier to distinguish. On the other hand, loose fabric may allow the fabric to flap against the post or other fence parts causing increased noise and increased nuisance alarms.

The rate of nuisance alarms for fence disturbance sensors can be affected by many factors including rain, hail, melting ice, wind, wind-blown debris, lightning, and the physical condition of the fence on which they are mounted. The magnitude of the effects of each of these sources depends on the type of sensor, the installation method, sensor maintenance practices, and the adjusted sensitivity or other settings.

It is also important that some fence disturbance sensors are not used near an electrical high-power line. Radios may also have an impact on the geophonic type of fence disturbance sensors. Immunity to these sources is one advantage of using fiber-optic technologies.

Since fiber-optic cable senses vibrations, nuisance alarm sources can be similar to those of seismic sensors. Anything causing a noise or vibration in the fence structure or fabric will likely be detected by the sensor. New digital processing algorithms incorporated into most processors have some capabilities to filter out extraneous noise such as that produced by wind or other vibrations. So, as with other fence disturbance sensors, the general condition of the fence is very important for the proper performance of this technology.

3.6.4 Characteristics and Applications

The following describes the characteristics and applications of fence disturbance sensors:

- Some cable types will provide the location along the fence disturbance sensor at which a detection occurred, which allows a quicker and more accurate assessment of an alarm.

- Most fence disturbance sensors are relatively inexpensive to purchase and inexpensive to install. This is particularly true since most facilities already have some sort of fence installed as a boundary marker.

- A fence disturbance type of sensor can be applied with excellent reliability to protect an asset located indoors. One example of this is a sensor applied to a cage that is built around the asset.

- An adversary who climbs or cuts a fence very slowly may be able to spoof a fence disturbance sensor; however, this can be much harder to do than it sounds. This scenario may be exacerbated if the adversary knows the event count and time constraints for alarm generation.

- A typical stand-alone fence disturbance sensor installation can be bridged over or trenched under.

- A fence disturbance sensor should not be used on the outer fence or as a facility's only line of detection.

- Typical rigid conduit installed on the fence will expand and contract with temperature changes, which will cause nuisance alarms.

- Fence disturbance sensors (except for fiber-optic technologies) should not be used near a high-power line. Radios may also have an impact on the geophonic type of fence disturbance sensors. Careful testing and evaluation should be conducted at the site before installation of the sensors to determine the frequencies and wattages that may be problematic.

3.6.5 Installation Criteria

It is possible to install a fence disturbance sensor on an existing chain-link fence; however, there is generally too much "noise" if the fence was not installed with the original goal of supporting such a sensor, and many adjustments will be required over time to achieve acceptable sensor performance. The following guidelines are suggested for preparing and evaluating a security fence for compatibility with fence-mounted intrusion detection sensors designed to sense motion, displacement, and/or acceleration forces resulting from an intrusion attempt:

- The fabric of the fence should be located on the unprotected side of the posts.

- The fence fabric should be stretched to allow not more than 2.5 inches of deflection when a 30-pound perpendicular load is applied at the center of the panel.

- The fence posts should not deflect more than 1.9 centimeters (.75 inches) when a 50-pound perpendicular load is applied at a height of 1.5 meters (5 feet) above the base of the post.

- The fence fabric should not be embedded in a concrete or otherwise solid sill. This will preclude later tightening of the fabric if necessary.

- The fence should not have a top rail. Top rails reduce sensor performance while aiding the climbing adversary. However, the fence should have a bottom rail.

- The fence should not have outriggers that tend to be noisy and cause unwanted nuisance alarms.

- Tension wires should be used to provide for the stabilization of the fence fabric.

- The top of the fence posts should be approximately 4 inches below the top of the fabric to prevent an adversary from using the post as a bridging aid or leaning a ladder against it.

- Wire ties should be used to secure the fence fabric to the structural members of the fence (i.e., posts, braces, or tension wires). The wire ties should be 9-gauge or thicker zinc-coated steel. Refer to Figure 33.

- The fence fabric should be secured with wire ties approximately every 12 inches.

- The wire ties should form a 540-degree tightened loop that is secure enough to prevent movement between the fence fabric and the wire tie.

Outside protected area

Inside protected area

Figure 33: Example of security fence wire ties.

Note that some fence disturbance sensors can accept input from an anemometer (windspeed indicator) so that the processing of events can be changed when windspeed exceeds a certain speed. Care should be taken when using this type of system as it can provide an automatically increasing attenuation of the sensitivity of the system as the wind increases. The wind can increase to a point where the sensor is no longer providing intrusion detection, unbeknownst to the operator. It is preferable to allow the operator to put the sensor into access if the nuisance alarms increase to an intolerable level and compensatory measures can be implemented as necessary.

Certain fence disturbance sensors will allow the sensitivity of each meter, or so, of the cable to be adjusted independently along the length of the cable to compensate for variations in the fence such as corners, tension posts, and gates. This flexibility also allows zones to be user defined in the software of the processor, potentially reducing the number of processors necessary.

Signs should not be mounted on a fence that supports a fence disturbance sensor. During windy conditions, the sign is likely to be a source of noise or provide a larger surface and thereby increase the fence movement and noise and, consequently, increase the number of nuisance alarms. If a sign must be mounted on a fence supporting a fence disturbance sensor, it should be attached with twisted metallic wire ties to secure the sign from rattling.

If chains are used anywhere along a fence that supports a fence disturbance sensor, such as to secure a gate, the chain should be rubber coated and as short as possible.

Tree branches, weeds, and other objects that could cause mechanical disturbance of the fence should be removed. Gates, barbed wire, and outriggers (when used) should be mechanically sound and, where appropriate, firmly attached to other parts of the fence. Such precautions prevent clatter or mechanical disturbances within the fence itself.

3.6.6 Testing

Three types of testing need to be performed during the life of a sensor: acceptance testing, performance testing, and operability testing.

3.6.6.1 Acceptance Testing

When a sensor is first installed, it should be tested in order to formally "accept" the sensor as part of the physical protection system. Acceptance testing consists of two parts:

(1) A **physical inspection** to ensure that the sensor is installed properly, as follows:

- Verify that the installation matches the installation drawings, which should follow the guidance provided by the manufacturer.
- Verify sector intersection spacing.
- Verify that signal and power wires are routed in conduit.
- Verify proper power levels, both voltage and amperage.
- Verify correct wire connections.

(2) A **performance test** to establish and document the level of performance (see next section).

3.6.6.2 Performance Testing

Performance tests should be designed to verify the level of performance of the sensor through its range of intended function. Fence disturbance sensors are designed to detect an adversary attempting to climb over or cut through the fence.

The following are examples of when a performance test should be conducted.

- A processor is replaced.

- Sensor cables are replaced or repaired.

- An adjustment that can affect sensitivity is made to the processor parameters.

- Configuration of the installed sensor cables is changed.

- There has been a change to the sensor or in the near vicinity of the sensor that might affect the detection sensitivity.

This test should include a visual inspection of the sensor and of the general area where the sensor is installed. The testers should refer to the maintenance section and perform the required routine maintenance.

All relevant processor readings and parameter settings that can affect the detection capabilities or the nuisance alarm rate should be recorded. Each manufacturer will have different sensor settings. The test should involve comparison of the current readings to the last recorded readings to ensure that they fall within the manufacturer's specifications.

To test the detection capabilities against a climb, the tester will actually climb the fence fabric. To preserve the integrity of the fence fabric, test devices are available that provide a consistent, simulated cut disturbance to the fence without damaging the fence fabric. (Refer to Figure 34.)

Figure 34: An example of a fence-cut simulation tool in use.

Sufficient testing should be performed to provide the probability of sensing (P_S) at the confidence level defined for the facility. For example, a "30/30" test methodology and binomial distribution may be used to yield a 90-percent P_S performance at a 95-percent confidence level. Using this methodology, if the sensor alarms 30 times when subjected to 30 attempted intrusions, the P_S value is equal to 0.90 at a 95-percent confidence level based on the values derived from a binomial reliability table.

Performance testing of fence disturbance sensors should include the following procedures. Performance tests should be conducted at 30 equidistant points along the length of the detection zone/sector. One or more test trials should be performed for each of the 30 test locations. Particular attention should be directed to locations where sensitivity might vary if the installation configuration varies from the norm, such as at corners, sector overlaps, or where fence structure is more significant.

3.6.6.2.1 Procedure 1: Conduct Climb Tests

Conduct 30 climb tests within the detection zone/sector to verify P_S = 0.90 at a 95-percent confidence level.

Each sector should be tested 30 times at points equidistant along the length of the sector. For this procedure, the adversary test subject should attempt to climb the fabric from the outer (unprotected) side of the fence. (Refer to Figure 35.) If climb testing of the fence is required from the inner side, do not use the fence support posts to aid in climbing. A failure is recorded if the adversary test subject can reach the top of the fence and start to go over without an alarm being generated.

Figure 35: Test climbing the fence from the outer (unprotected) side.

3.6.6.2.2 Procedure 2: Conduct Simulated Cut Tests

Using the simulated cut device, conduct 30 simulated cut tests within the detection zone/sector to verify PS = 0.90 at a 95-percent confidence level.

Each sector should be tested 30 times at points equidistant along the length of the sector. For each trial, use the cut simulation tool to strike the fence (releasing the plunger from the middle notch) X1 times at a point either 12 inches from the ground, 12 inches from the top of the fence, or at a point in the middle of the fence fabric. Ensure that the test device is firmly pressed against the fabric when releasing the plunger. The strikes should be at approximately X second intervals. A failure shall be recorded if the tester can strike the fence X times within the designated timeframe without generating an alarm.

3.6.6.2.3 Procedure 3: Verify Processor Enclosure Tamper Alarm

Test the tamper alarm of the processor enclosure by slowly opening the door to the enclosure. A failure shall be recorded if an alarm is not received or if the cover can be opened more than 1 inch before an alarm is received.

---

[1] X = The numbers for the amount of times needed to strike the fence and the required time interval to generate an alarm which are determined by the site and are usually classified.

3.6.6.2.4 <u>Resettling Time</u>

When following the performance test procedures, allow sufficient time for the system under test to stabilize or reset between each individual test trial.

3.6.6.2.5 <u>Verification of Failure to Alarm</u>

If an alarm is not received for a given trial, two more attempts may be made at the same location using the same test method. The trial should be considered a valid detection and testing can proceed **only if** both of the additional trials result in successful alarms. If either additional trial results in a failure, testing of the system should cease and corrective action should be taken. Following required maintenance or recalibration, the complete performance test procedure should be conducted again using the same test methodology at all test locations (if possible) to verify acceptable sensor performance. If sensing performance does not meet requirements after all reasonable corrective actions have been attempted, the test should be annotated as a failure.

3.6.6.3 Operability Testing

The operability test is conducted by causing a disturbance to the fence fabric within the zone to be tested. The disturbance can be caused by striking the fence with some object or simply shaking the fence until an alarm is received at the alarm monitoring station. The striking or shaking force to be exerted during the operability test is usually outlined in manufacturer's specifications and is derived from a reasonable representation of the force that the system would encounter and should detect if a human were attempting to circumvent the system by climbing or cutting.

If no alarm is received, a maintenance request should be immediately generated and implementation of compensatory measures should be considered based on site-specific requirements.

3.6.7 Maintenance

At least twice a year and after any major storms, each section of a fence disturbance sensor system should be carefully examined to make certain that no parts of the sensor system have become loose or the integrity of the fence has not been compromised. The inspection should include the following:

- Ensure that wire ties for the sensor cable have not broken or become loose.

- Ensure that all fence fabric wire ties remain tight and are securely attached to the fence structure.

- Inspect for any debris that has collected against the fence and remove this debris.

- Inspect any signs that may be mounted on the fence fabric and ensure that they are securely fastened.

- Ensure that any splices or terminations in the sensor cable remain sealed and securely attached to the fence.

4. INTERIOR INTRUSION DETECTION SENSORS

4.1 Balanced Magnetic Switch

4.1.1 Principles of Operation

The magnetic switch is the most ubiquitous security device in the world, used for both door and window protection. (Refer to Figure 36). It consists of a magnet, which is installed on a door or window, and a switch unit, which is installed on the frame. When the door or window is in the closed position, the created circuit is closed; when the door or window is opened, the circuit is open, which causes an alarm to be initiated.

Figure 36: An example of the application of a magnetic switch for door protection.

At a cost of approximately $10, the simple magnetic switch device is used in most residential and small business security systems. Unfortunately, little knowledge is required to defeat (bypass or spoof) a simple magnetic switch. In response, the balanced magnetic switch (BMS) was developed more than 30 years ago. The BMS requires a great deal more skill to defeat than a simple magnetic switch.

A BMS employs a magnet in both the switch and magnet units. (Refer to Figure 37). The switch unit, which contains a magnetic reed switch, a bias magnet, and tamper/supervisory circuitry, is mounted on the stationary part of the door or window unit. The component containing the larger permanent magnet is mounted on the movable part of the door or window, adjacent to the switch unit installed on the frame when the door is closed.

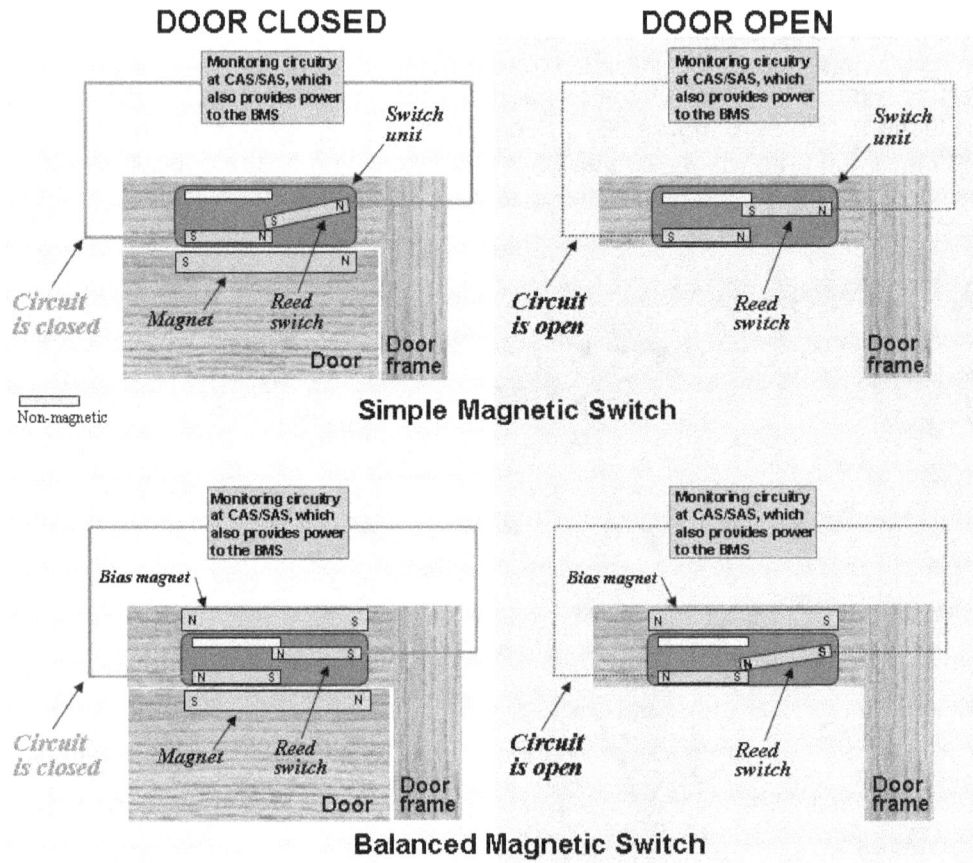

Figure 37: A schematic of a simple versus balanced magnetic switch.

With the door or window closed, the magnets are adjusted to create a magnetic loop, causing the reed switch to experience a magnetic field of essentially zero. When the door or window is opened and the magnetic field is removed, the contacts will separate and trigger an alarm indicating a security breach. In some models, this magnetic field is accomplished by adjusting the bias magnet; in other models the adjustment is made by varying the position of the magnetic unit with respect to the switch unit.

Any action that causes the magnetic field to become unbalanced, such as opening the door or window, results in the transfer of the reed switch to the "closed" position and an alarm output. The same result is obtained if an external magnet is brought into the vicinity of the BMS, thus changing the magnetic field.

A newer BMS codes multiple magnets in each unit; thus, each BMS is matched and the defeat of the BMS is nearly impossible. However, fixing a switch in a BMS that has a coded configuration requires that the BMS be completely replaced with another matched pair.

A BMS is a passive, visible, point-detection sensor.

4.1.2 Types of Balanced Magnetic Switch Sensors

BMS sensors have a variety of designs. Some use multiple magnets; some have internal electromagnets for self-testing. Manufacturers have reduced the vulnerability of magnetic switches to external magnetic fields through the following measures:

- Using multiple magnets, various magnetic orientations, and magnetic shielding, such as Mumetal®

- Creating standoff distances

- Adding magnetic tamper indicators

The newest switches on the market have very narrowly defined magnetic field paths, making them almost immune to external magnets. All provide a high level of protection for access points such as windows and doors. The type of BMS a facility chooses should be based on providing reliable functionality within the environment and the goals of the physical protection program.

4.1.3 Sources of Nuisance Alarms

BMS sensors are very reliable when installed correctly on a properly installed door with hardware that is in good condition. Nuisance alarms are almost never caused by the BMS alone.

Most nuisance alarms generated by a BMS can be attributed to the poor condition of the door or its hardware. Most common nuisance alarms are caused by a worn or out-of-adjustment latch that occurs because of excessive door movement or play. Nuisance alarms can also be caused by excessively worn door hinges or an improperly installed BMS that causes misalignment of the switch unit and magnet unit.

When a BMS is used on large rollup doors, nuisance alarms are usually caused by slight misalignment of the doors. It is difficult over time to keep this type of door maintained in alignment for proper sensor operation. Extreme weather conditions that cause excessive movement of a door, window, or access portal can also increase the nuisance alarm rate of a BMS.

4.1.4 Installation Criteria

The switch assembly of a balanced magnetic sensor is mounted on the inside of the fixed surface and the magnetic assembly is mounted near the top of the movable surface near the edge that is on the opposite side of the hinge. This mounting allows for maximum detection of movement.

A BMS should always be installed on the secure side of the door.

If the BMS is installed on a recessed door or outward opening door, a spacer will be needed to line up the switch and magnet units. If the door frame is steel, a nonferrous (aluminum or plexiglass) spacer should be installed between the door frame and the switch unit to prevent interference with switch operation. Likewise, if the magnet is installed directly on a steel door, it should have the same type of spacer. A spacer made from ½-inch-thick plexiglass has worked well in many installations. Some manufacturers state that their switch compensates for the effects of steel. It is best to consult the manufacturer to verify the need for a spacer.

The wiring from the BMS should be protected. Installing the wiring in a conduit from the sensor switch enclosure all the way to the alarm data-gathering or multiplexer panel will provide protection for the alarm wiring. Materials with high magnetic permeability, such as Mumetal®, are preferable for the shielding. However, steel can also be used.

Sensor electronics enclosures should have tamper switches. Line supervision is the means for monitoring the communication link between a sensor and the alarm control center. Use of supervised lines between the sensor and host alarm system, as well as continuously monitored sensor tamper switches, will help protect against adversary attacks on communication links and sensor electronics enclosures.

4.1.5 Characteristics and Applications

A BMS is passive and visible and detects boundary penetration such as a door or window being opened. These switches are manufactured in different sizes and shapes and achieve different performance levels, depending on the manufacturer and the model.

A BMS is a mature technology that is subject to few (if any) nuisance alarms, provided that the door, door frame, and door hardware are in good condition and that the BMS was installed properly.

An externally introduced magnetic field has the possibility of defeating a BMS. BMS sensors with multiple magnets and reed switches will be much more difficult to defeat by this method. If a door magnet can be removed without detection, it may be possible to compromise the BMS. These sensors provide protection only if the intruder opens the door or window for entry. If the intruder cuts through the door, cuts through the wall next to the door, or breaks the window pane, the BMS will be bypassed. Consideration should be given to bolstering the resistance to adversary penetration of these potential pathways.

High-voltage discharges from lightning, power surges, or stun guns can permanently weld reed switch magnetic contacts in a failed (closed) position, making the system useless when it is armed. If the metal contacts are welded shut, it will indicate a secure state even when the system is breached.

4.1.6 Testing

A regular program of testing sensors is imperative for maintaining them in optimal operating order.

4.1.6.1 Acceptance Testing

When a BMS sensor is first installed, it should be tested in order to formally "accept" the sensor as part of the physical protection system. Acceptance testing consists of two parts:

(1)   A **physical inspection** to ensure that the sensor is installed properly consists of the following:

- Verify that the installation matches the installation drawings, which should follow the guidance provided by the manufacturer.
- Verify that signal and power wires are routed in the conduit.
- Verify proper power levels (voltage and amperage).
- Verify correct wire connections.

(2)   A **performance test** to establish and document the level of performance (see next section).

4.1.6.2   Performance Testing

If a BMS has had an unexplained number of nuisance alarms or if the BMS has ever failed to generate an alarm during a daily walkthrough test, troubleshooting and/or repair will be required; after this repair, a formal performance test should be run. Also, operation of the sensor tamper and communication to the alarm stations are verified during testing.

Three basic tests constitute a performance test for a BMS:

(1)   Evaluate the response of the switch when an externally introduced magnetic field is produced via a foreign magnet during testing. (Refer to Figure 38.)

(2)   Establish that the BMS detects a door opening within a specified distance. A commonly used requirement is that a BMS shall generate an alarm when the leading edge of the door has been moved 1 inch or more from the fully closed position.

(3)   Establish that the BMS does not detect a door opening within a smaller specified distance. A commonly used requirement is that a BMS shall not initiate an alarm for door movement of ½ inch or less. The importance of the condition of the door and its associated hardware cannot be overemphasized. If the door is not properly installed and maintained, the BMS effectiveness will be degraded.

The history of nuisance alarms and false alarms should be reviewed at this time as well. Establishing specific values for false alarm rates helps the operator determine when a sensor should be reported to maintenance personnel.

4.1.6.3   Operability Testing

The objective of the operability test is to verify that the sensor is operational and that the correct alarm signal is received and displayed at the alarm stations.

The BMS operational test is simple and is performed by opening the door and verifying with the alarm station operator that an alarm was received at that particular door location. Then another test is conducted to verify with the operator that the sensor returns to the secure state and remains secure when the door is closed, latched, and pushed back and forth, given that there is some play in the latch.

Figure 38: Demonstration of introducing an external magnetic field to a BMS. Note that this procedure is much more difficult to accomplish than the photos would suggest.

The door hardware should be in good condition. The door should open and close smoothly without rubbing or scraping on the door frame. It should latch easily and the amount of play in the latch should be minimal.

In addition to scheduled operability tests, operability testing should be performed when a protected location is placed into a secure condition from an unsecure condition (i.e., entrances locked and alarms reactivated).

4.1.7 Maintenance

At a minimum, maintenance of a BMS should be performed every 6 months. The following should be verified during maintenance:

- Verify tamper operation by accessing electronics enclosures and disconnecting communicating links of the sensors and through successful communication to the alarms stations.

- Verify tamper operation or alarm when a foreign magnetic field is introduced to the sensor.

- Verify that an alarm occurs within a specific door movement distance, which is typically before the leading edge of the door has moved 1 inch.

- Verify that no alarms occur during any slight movement of the door when it is latched.

- Verify acceptable conditions of electrical power and communication lines.

4.2 Interior Microwave Sensors

4.2.1 Principles of Operation

Interior microwave sensors are active volumetric sensors and are typically monostatic, employing a single antenna for both the transmit and receive functions; all components are enclosed in a single housing. (Refer to Figure 39.) Microwave sensors emit an energy field.

Motion within an area protected by a microwave will cause changes to the microwave energy, and these changes are a type of Doppler frequency shift. A person or other object moving within the microwave energy field will cause minute changes in the frequency of the microwave. As the sensor "knows" the frequency at which it is transmitting, when it receives reflected energy at a slightly different frequency, it will process the difference between the frequencies. An alarm will be generated if the frequency difference exceeds a preset threshold.

Microwave sensors typically operate in the X band radiofrequency region (7 to 11 gigahertz) with low power output that is approximately 5 to 10 milliwatts. Figure 40 illustrates typical sensor coverage patterns. The size and shape of this pattern can vary significantly, depending on the characteristics and configuration of the microwave antenna used in the sensor design, although most interior monostatic microwave sensors have a detection pattern that ranges from approximately 9 meters (approx. 29.5 feet) up to 30 meters (approx. 98.4 feet) in length. The shape of the detection zone is governed by the design of the antenna and is roughly similar to an elongated balloon or a cigar. The antenna is usually a microwave horn but may be a printed circuit planar or phased-array antenna.

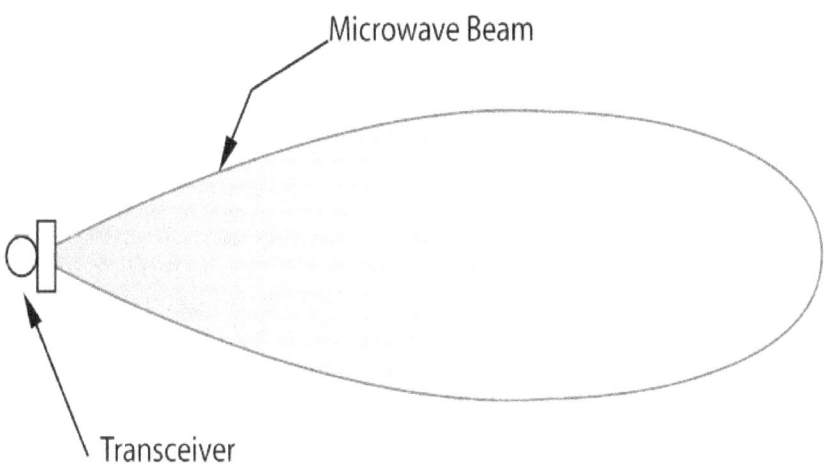

Figure 39: A common interior microwave antenna propagation pattern.

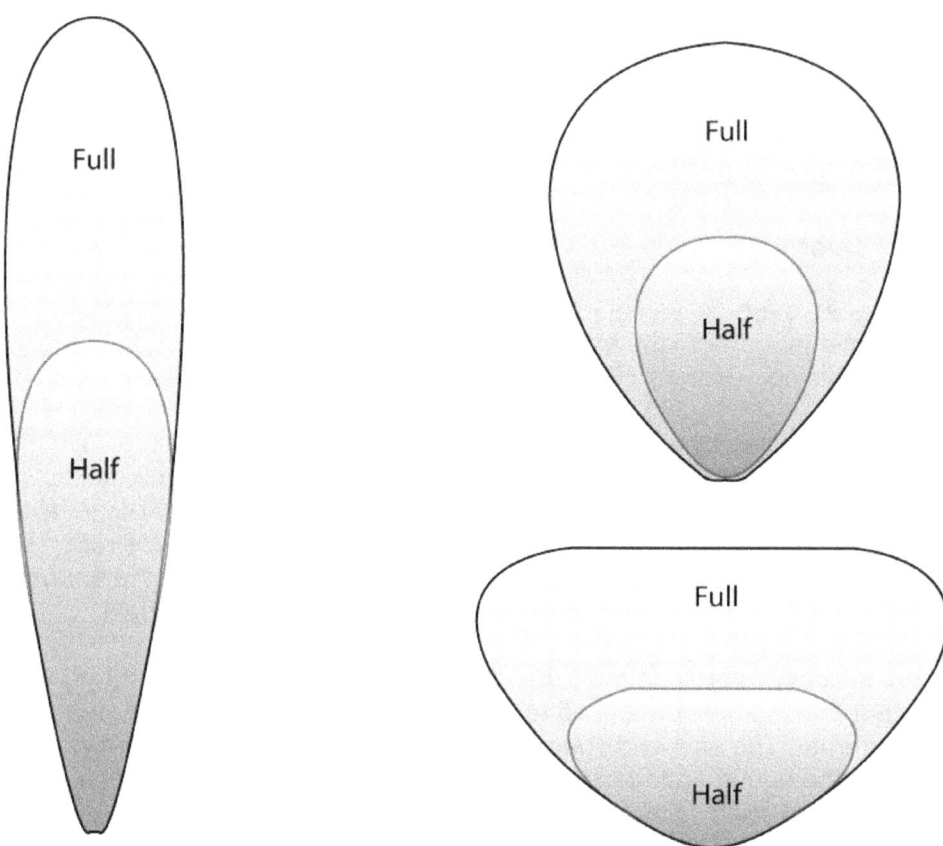

Figure 40: Examples of different microwave antenna propagation patterns; the half and full envelopes indicate half-power and full-power settings.

Optimal detection for microwave sensors is achieved when the target is moving toward or away from the sensor—not across the detection zone. Therefore, placement of microwave sensors should be such that the adversary will be necessarily forced to move towards or away from the sensor to accomplish the adversary's objective.

4.2.2 Types of Interior Microwave Sensors

The two basic types of microwave sensors are monostatic sensors, which have the transmitter and receiver encased within a single housing unit, and bistatic sensors, in which the transmitter and receiver are two separate units creating a detection zone between them. A bistatic system can cover a larger area and would typically be used if more than one sensor is required for the area being covered, but is more commonly applied in exterior applications.

4.2.3 Sources of Nuisance Alarms

Because of the high frequencies of microwave sensors, the signal/sensor is not affected by moving air, changes in temperature, or humidity. However, the high frequency allows the signal to pass through standard walls, glass, sheetrock and wood, which can cause nuisance alarms to be generated by movement adjacent to, but outside, the detection area. If a microwave sensor is installed in a room made from light construction materials and the detection area of that microwave is larger than the room, nuisance alarms will occur because of movement outside the room. The structural materials and the thickness that a particular microwave can penetrate will vary, based on the device's manufacturer, model, and frequency. (See Figure 41.)

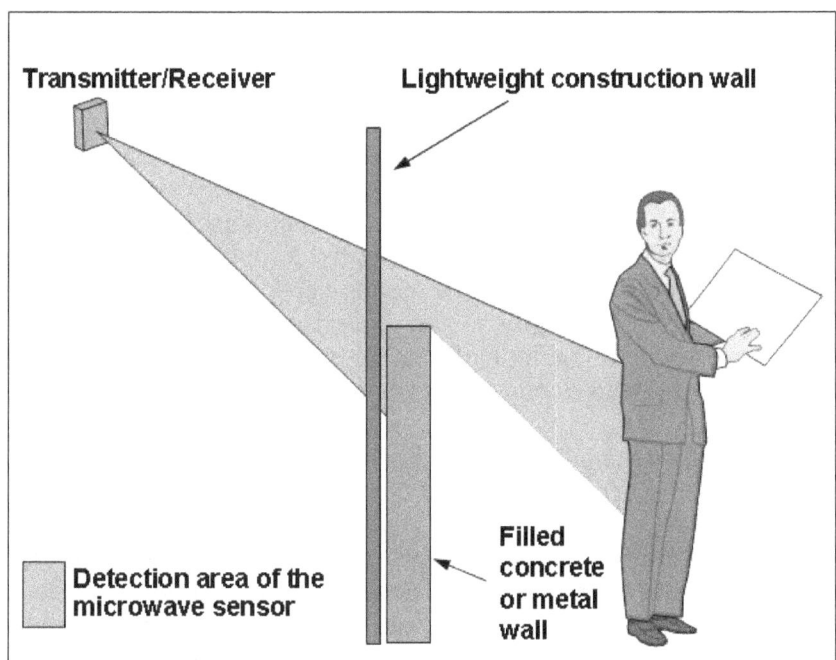

Figure 41: A microwave sensor can trigger nuisance alarms because of its ability to transmit through some light-weight construction materials.

Nuisance alarms can be generated by fans, rodents, pets, or equipment. A fan that is not located in the room of primary detection is also a possible source. If a fan is located within a ventilation duct, reflections from the moving fan blades can be detected by microwave energy traveling through the duct. Another source of nuisance alarms for microwaves is plastic drainpipes located behind dry wall. Water draining through the pipes can cause alarms.

Some microwave sensors can be triggered by fluorescent lights. The gas within fluorescent lights is a reflector of microwave energy when ionized. The light actually flickers at the power line frequency, which the sensor perceives as motion. If nearby microwave sensors generate nuisance alarms, a metallic screen mesh (also known as a Faraday cage) can be installed over the lights to prevent the microwave energy from passing through.

Electromagnetic sources close to the microwave frequency are another possibility. In fact, if more than one microwave sensor is installed in the same area, they can potentially interfere with each other and cause nuisance alarms. Fortunately, microwave sensors are available that allow the user to select from several different frequencies. More than one microwave in the same area will require different frequencies.

Finally, signals reflected off metal objects, such as filing cabinets, trash cans, and electrical boxes, can "extend" sensor coverage to areas not intended to be covered, thus creating the potential for nuisance alarms.

4.2.4 Characteristics and Applications

Microwave sensors are most sensitive and effective when installed so that an adversary would necessarily walk toward or away from the sensor.

Microwave sensors can be used to effectively monitor interior confined spaces such as vaults, special storage areas, hallways, and service passageways. They can also serve as an early warning alert of intruders approaching a door or wall. In situations where a well-defined area of coverage is needed, the use of monostatic microwave sensors is appropriate.

To further enhance detection, a facility can install a complementary sensor, such as a passive infrared (PIR) sensor. (A PIR sensor is considered to be complementary to a microwave sensor because a PIR senses best when an adversary moves across the zone of detection, unlike the microwave.) The use of a complementary system provides a second line of defense and provides security personnel with additional information to help them accurately assess an alarm and discriminate actual or potential penetrations from nuisance events.

Microwave sensors are least sensitive if installed such that an adversary would be able to limit his or her movements to paths across the detection pattern.

The special properties of microwave beams allow them to penetrate most types of surfaces (metal is not one of them). Because of this, it is possible for a microwave to detect motion in an area where detection is not desirable and not detect motion where it is desirable. For example, a large metal filing cabinet in the area of detection will shield the area behind it. Objects such as these create "dead zones," areas where the sensor cannot detect motion, thereby creating a hiding place for a potential adversary. On the other hand, because the beam can penetrate walls, the sensor may detect motion behind a wall in another room.

As microwave sensors are extremely sensitive to motion, they are also prone to nuisance alarms. Objects being moved by air currents generated by the heating, ventilation, and air conditioning (HVAC) or fans may trigger alarms. Even fluorescent lighting, which emits detectable light particles, may trigger a false alarm.

Since microwave sensors operate in the high-frequency spectrum (X band), close association or proximity to other high-frequency signals can adversely affect their detection reliability. Areas that contain strong emitters of electric fields (radio transmitters) or magnetic fields (large electric motors or generators) can affect the ability of microwave sensors to function properly and should be avoided or compensated for by distinct signal separation or shielding. Self-generated signal reflection is a common problem caused by improper placement or mounting. Positioning the sensor externally and parallel to the wall rather than embedding it within the wall will aid in avoiding this problem.

Very slow movement by an intruder is harder for a microwave sensor to detect, though to actually defeat a microwave sensor is not easy. The speed required to bypass a sensor will depend on its make and model. Testing in the past has shown that some microwave sensors will still have some sensitivity against intruders moving at speeds as slow as 1 inch per second. The microwave sensor will detect any swaying of the body, including movement of the head, arms, or legs. For a successful defeat, an intruder will be required to be inside the detection area for a lengthy period to allow the time necessary to move this slowly and avoid detection, which thereby increases the chances that the intruder will be noticed.

Circumferential motion in a perfect arc, with no effective motion toward or away from the sensor, will not produce a Doppler shift, and hence, no detection will occur. This is, however, a very difficult movement for an intruder to accomplish correctly and subsequently avoid detection.

The three graphs in Figure 42 show the differences between the detection pattern shape and size with respect to test subject direction of movement into an area that is sensored by a microwave. The left graph shows the maximum detection pattern with the test subject walking directly toward the sensor. The right shows a small decrease in detection pattern size with walk testing parallel to the sensor center line. The top graph shows a much smaller detection pattern with walks that are parallel to the sensor face. This direction results with less of a Doppler frequency shift; the Doppler shift requires a sufficient amplitude change and duration time to cause an alarm. In practical terms, this means that the microwave transmitter sends out a known frequency and if a higher or lower frequency is returned to the receiver, the target is moving closer or further away from the sensor.

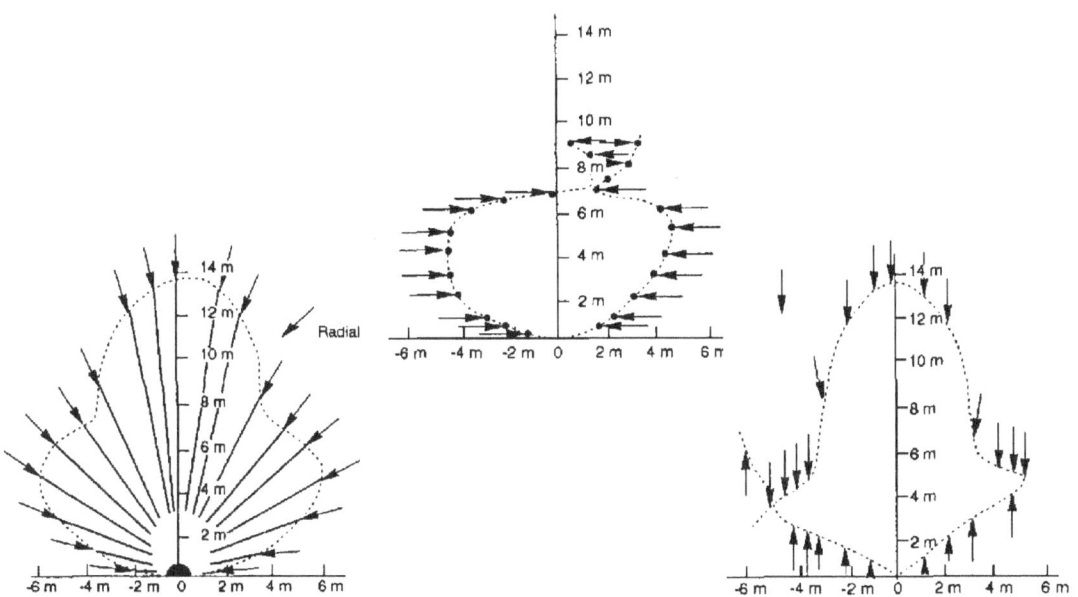

Figure 42: Differences in detection patterns occur for walk tests that are performed from a variety of orientations and from different directions.

4.2.5 Installation Criteria

Microwave sensors should ideally be mounted near the ceiling or directly on the ceiling. A rigid and stable mounting assembly should be used. The actual location of the sensor (ceiling, corner, wall, etc.) will depend on the particular sensor being used, as well as the area or target it is intended to protect.

Care should be taken when surveying the area to be protected to note any object that may degrade the detection capability of the sensor (metal filing cabinets, fans, air conditioner vents, etc.). Because microwave energy is difficult to constrain, special care should also be taken when locating and directing the energy within the area requiring detection. A protected volume surrounded by masonry or metal construction confines microwave energy and prevents detection outside the protected volume, thus preventing one common source of nuisance alarms.

Fluorescent lights located in the sensor detection envelope, especially at distances of less than 3 meters (about 10 feet), may cause low-frequency Doppler shifts, originating with reflections from the ionized gas within the fluorescent tubes. Blocking the line-of-sight path, by using either a metal mesh of 0.6 centimeters (0.25 inches) mesh or a radiofrequency absorber, eliminates such signal interference.

4.2.6 Sensor Testing

A regular program of testing sensors is imperative for maintaining them in optimal operating order. Three types of testing need to be performed at different times in the life of a sensor: acceptance testing, performance testing, and operability testing.

4.2.6.1 Acceptance Testing

When a sensor is first installed, it should be tested in order to formally "accept" the sensor as part of the facility's physical protection system. Acceptance testing consists of two parts:

(1) A **physical inspection** to ensure that the sensor is installed properly consists of the following:

- Verify that the installation matches the installation drawings, which should follow the guidance provided by the manufacturer.
- Verify that signal and power wires are routed in the conduit.
- Verify proper power levels (voltage and amperage).
- Verify correct wire connections.

(2) A **performance test** to establish and document the level of performance. Refer to the following section for details.

4.2.6.2 Performance Testing

Performance tests (refer to Figure 42) are designed to verify the level of performance of each microwave sensor through the range of intended function. This testing will verify the manufacturer's published detection pattern or will establish the actual detection pattern.

This test should include a visual inspection of the sensor and of the general area where the sensor is installed. Prudent routine maintenance should be performed according to the maintenance section.

All relevant processor readings should be recorded, and new readings should be compared to the last recorded readings. As in all test situations, the area under test should be maintained under visual observation by a member of the site security force, or a member of the site security force should actually conduct the test. For each sensor, the test should, where possible, do the following:

- Ensure that the system meets the manufacturer's specifications and recommended detection probability.

- Verify that no disabling dead spots exist in the zone of protection.

- Verify that line supervision and tamper-indication alarms in both the access and secure modes are functional.

- Verify that both line supervision and tamper-indication alarms are received in the alarm station as appropriate.

Records of initial testing capabilities, equipment sensitivity setting, or voltage outputs should be maintained so that deterioration in equipment capability can be monitored. Walk tests should be performed for all areas covered by the microwave sensor, and compared with the results of the acceptance test to check for any degradation in the coverage of the sensor.

4.2.6.2.1 Radial Path Testing

The following instructions describe the walk test to be conducted along the radial (parallel to the common center of the detection zone) paths (refer to Figure 43), which is the most effective detection approach against a microwave.

(1) Start outside of the published detection area in front of the sensor, and walk at 1 foot per second along the first radial path.

(2) Stop when an alarm occurs and mark that position.

(3) Return to the start point, wait 30 seconds for the sensor to reset, and repeat the walk test along the same path.

(4) Repeat testing on that path until the required number of tests is completed. Multiple tests along each test line path are required to establish a P_D (probability of detection). As an example, to establish that a sensor has a minimum P_D of 90 percent at a confidence level of 95 percent, the sensor would have to pass 29 out of 30 tests.

(5) Perform Steps 1 through 4 for the remaining radial paths.

4.2.6.2.2 Tangential Path Testing

The following instructions describe the walk test to be conducted along the tangential (lateral or perpendicular to the common center of the detection zone) paths (refer to Figure 43), which are the least effective detection approach against a microwave.

(1) Start outside of the published detection area on one side, and walk at 1 foot per second along the first tangential path.

(2) Stop when an alarm occurs and mark that position.

(3) Return to the starting point, wait 30 seconds for the sensor to reset, and repeat the walk test along the same path.

(4) Repeat testing on that path until the required number of tests is completed. Multiple tests along each test line path are required to establish a P_D (probability of detection). As an example, to establish that a sensor has a minimum P_D of 90 percent at a confidence level of 95 percent, the sensor would have to pass 29 out of 30 tests.

(5) Perform the above tests on remaining paths.

(6)   Repeat Steps 1 through 5, starting from the other side of the detection area.

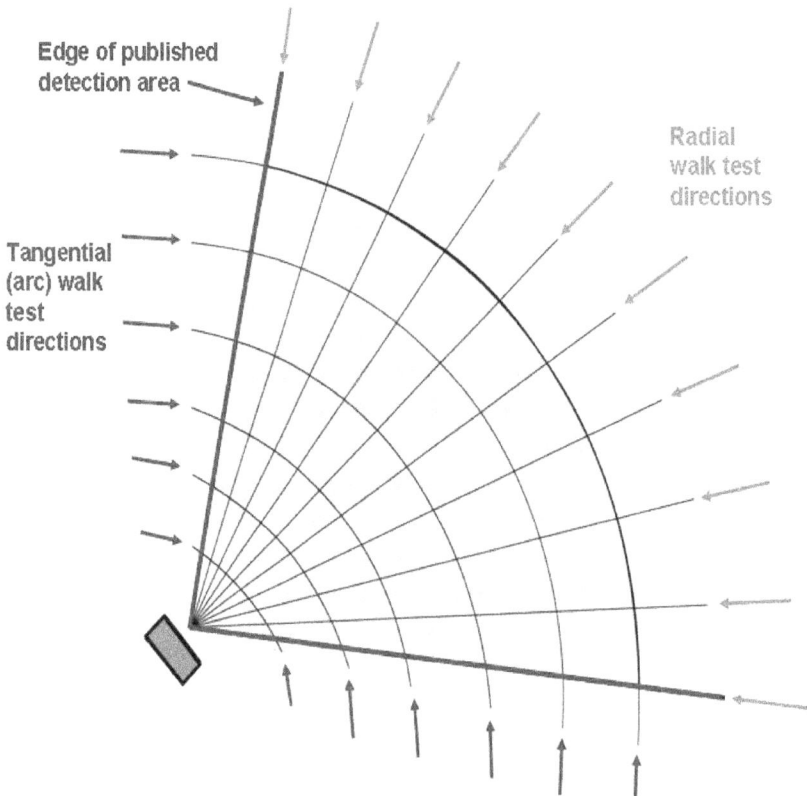

Figure 43: Recommended walk test paths and directions for the performance testing of a microwave sensor.

4.2.6.2.3 Slow Walk Tests

Slow walk tests are conducted at speeds less than 0.5 feet per second. Most volumetric sensors such as the microwave will have a speed where detection capability decreases. If the potential to circumvent a system by crawling is a concern, crawl testing should be performed to obtain detection characteristics. Detection of a crawling person will likely be different than detection of a walking person.

If an equal number of tests for each approach is not possible, the penetration approach pattern that is most difficult to detect for a particular sensor should be attempted more frequently. The various paths should be tested in random order, which will reduce the possibility that environmental effects and other unknown factors are affecting the test results (i.e., detection or nondetection). Using a random sequence, there is less chance that the test results would be biased.

4.2.6.3   Operability Testing

Operability tests for these systems consist of simple walk tests. The testing individual walks through the expected detection zone of a sensor and confirms that the alarm has been received at the alarm display center. The testing individual should look for any evidence of damage to the sensor or tampering with the device.

4.2.7 Maintenance

A visual inspection of the installation should be performed quarterly and immediately after major maintenance to the building in the sensor area. Mounting brackets and hardware should be inspected for stability and corrosion. Frequent visual inspections ensure that no blocking objects have been moved into a position that would render the sensor inoperative. Periodic tests, in addition to the self-test invoked by the sensor or the system, ensure that the sensor is operating effectively. Standby batteries should be replaced on a conservative schedule. Every service call should be entered in a log to record the date, time, corrective action, and an assessment of the cause of the problem.

4.3 Passive Infrared

4.3.1 Principles of Operation

PIR sensors are the most commonly used volumetric sensor for interior applications. Many facilities use PIRs for the protection of the interior of rooms or particular areas of a room. (Refer to Figure 44.)

Figure 44: Detection pattern for a typical installation of a PIR sensor.

PIR sensors detect the electromagnetic radiated energy generated by sources that produce temperatures below that of visible light. PIR sensors do not emit any energy field into the area they are protecting and do not measure the amount of infrared energy. Rather, PIRs measure changes in thermal radiation. PIR sensors detect thermal radiation by sensing the change in contrast between a heat source and the ambient background temperature. They are considered to be a type of visible sensor because they are in plain view within the area; they also require a line of sight between the sensor and any target to be detected. The sensor detects intrusions as a function of the magnitude of the difference between the intruder's temperature and the background temperature.

Using parabolic mirrors or Fresnel lens optics, the infrared energy is focused on the detector chip in the sensor. Using either variety of lens, the detection pattern is subdivided into solid angular segments. (Refer to Figure 45.) As a person passes across the detection segments, each segment passed through will generate an increase or decrease in temperature, which will trigger an alarm. This infrared energy is detected by a thermopile or pyroelectric device and converted into an electrical signal. This signal is then processed by circuitry in the sensor which determines whether this constitutes an alarm. The electronic processing can be a count of the number of signal pulses over the detection threshold and will generate an alarm only when a specified number of pulses occurs within a certain time period. An alarm is annunciated when the difference between an intruder and the ambient background temperature reaches a predetermined value. On some sensors, this difference can be as small as 1 degree Celsius.

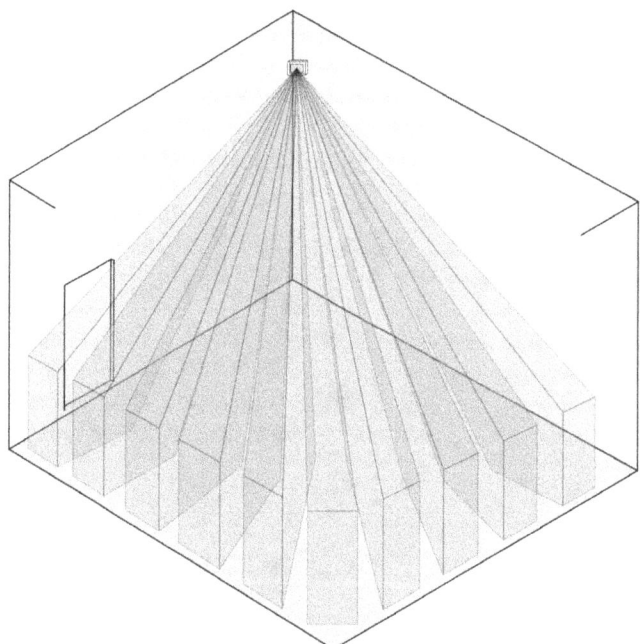

Figure 45: A PIR sensor's detection pattern is subdivided into solid angular segments; a specific number of segments must detect an anomaly before the sensor will signal an alarm condition.

While infrared radiation is invisible to the human eye, infrared radiation emitted by an object is directly related to its temperature. The infrared region lies between 0.75 and 1,000 micrometers. The human body radiates infrared energy in the 8 to 14 micrometer region.

PIR sensors continuously receive infrared energy from all objects within an area being protected. Ceilings, walls, floors, furniture, and other objects all emit infrared energy proportionate to their temperature and emissivity (emissivity defines how well an object absorbs and radiates infrared energy). A PIR will respond only to changes in the received infrared energy. The absorption and radiation of infrared energy depend on the composition of the surface of the object.

4.3.2 Types of Passive Infrared Sensors

By configuring the parabolic mirrors or Fresnel lens, a single conical field (referred to as a "curtain PIR"), a multiple segment field, or a hemispherical field of view can be generated, as shown in Figure 46 and Figure 47.

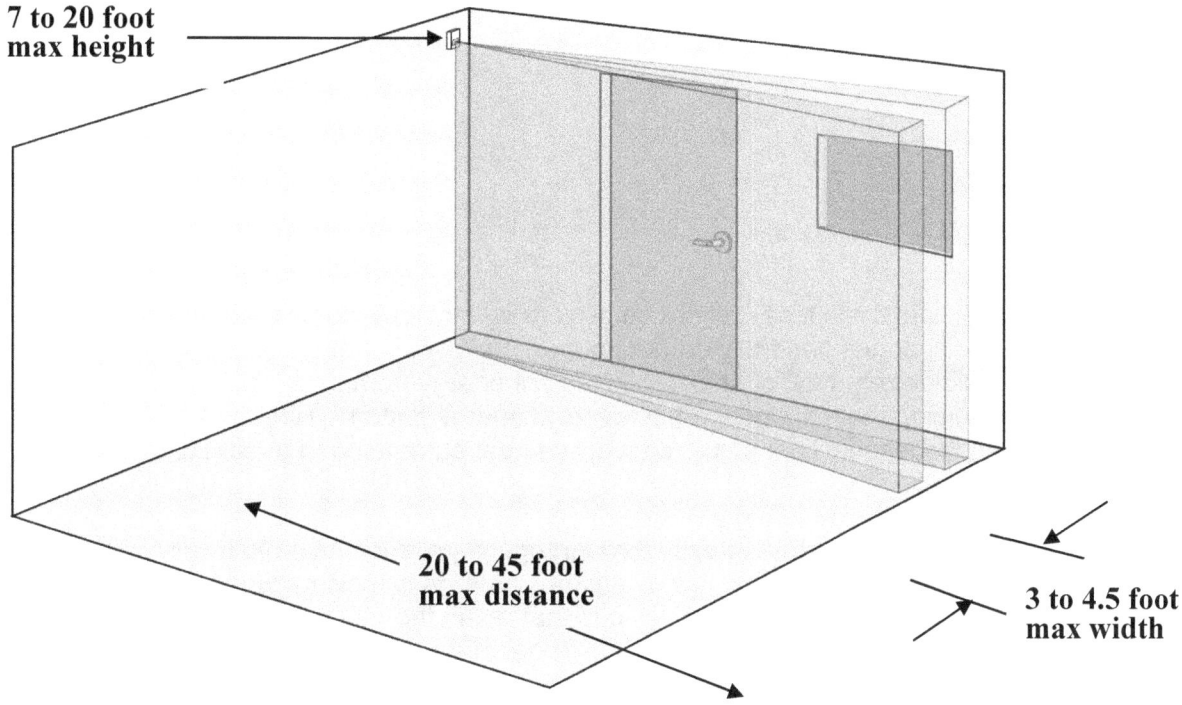

7 to 20 foot max height

20 to 45 foot max distance

3 to 4.5 foot max width

Figure 46: An example of a curtain PIR application.

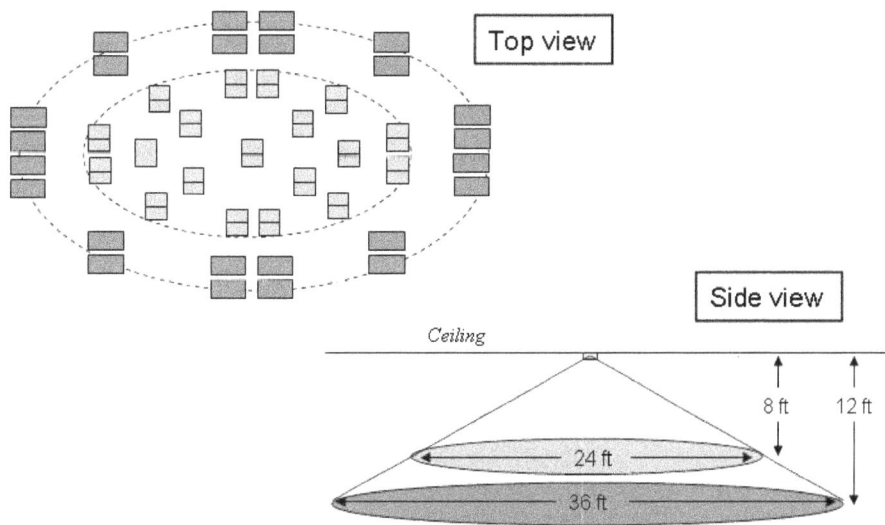

Top view

Side view

Ceiling

8 ft   12 ft

24 ft

36 ft

Figure 47: These two diagrams illustrate the detection pattern of a ceiling-mounted PIR sensor (top view and side view). This configuration is also known as the hemispherical pattern.

4.3.3 Sources of Nuisance Alarms in Passive Infrared Sensors

Any object causing an appropriate temperature differential can potentially generate a nuisance alarm in a PIR sensor. The required temperature differential can be caused by rapid changes in localized heating and cooling, which may be affected by the following:

- Sunlight

- Incandescent light bulbs

- Radiators

- Space heaters

- HVAC vents

- Hot pipes

In practice, nuisance alarms are less likely to be generated by localized heating and cooling because temperature changes generally do not happen rapidly. All hot spots that generate infrared energy should be removed or shielded. Radiant energy from such sources may produce thermal gradients that change the background energy pattern. Hot spots can be open heating elements, incandescent light bulbs, convective heat currents, and direct sunlight on windows, floors, and walls.

Sunlight can enter the protected area directly through openings, such as broken window panes, ventilation grids, and poorly fitted doors. Small animals or large insects moving in the PIR sensor field of view may also be detected. Devices that retain the required temperature differential and sway into the sensor field of view may generate nuisance alarms.

The vibration of a PIR sensor may trigger an alarm by causing a heat source to appear as if it were moving. Insects crawling on elements inside the sensor or condensation forming within the sensor may also cause nuisance alarms.

The detector elements in a PIR sensor can be subject to interference from various electromagnetic fields generated by electromagnetic devices, such as hand-held radios. However, infrared sensors are not generally subject to nuisance alarms caused by sound, electrical disturbances, or radio disturbances.

4.3.4 Characteristics and Applications

PIRs are installed so that the detection pattern covers the area or asset to be protected. This detection pattern can be pictured as a "searchlight beam" that gradually widens as the zone extends farther from the sensor with some segments being illuminated while others are not. This design characteristic allows the user to focus the beam on areas where detection is needed while ignoring other areas, such as known sources of false alarms.

Changes in the infrared signature of an object (including people) are most visible when the object moves laterally through the detector's range. Positioning a detector so that an intruder must walk across the detector's range is much more effective than positioning the detector so that an intruder would likely walk toward the detector.

The presence and/or location of a passive sensor is more difficult for an intruder to determine than an active sensor, putting the intruder at a disadvantage.

In environments where explosive vapors or explosive materials may be present, passive sensors are safer than active ones because no potentially explosion-initiating energy is emitted.

Multiple passive sensors can be placed in a volume without interfering with each other (interacting) because no signals are emitted.

Detection effectiveness is less than optimal for motion directly toward or away from the sensor.

Since the PIR is a line-of-sight detector, the field of view can be easily blocked by cubicle partitions or furniture.

Sources of rapid temperature changes can cause nuisance alarms. Sensitivity changes with the temperature of the detection area. If the ambient temperature is near the body temperature of an intruder, the intruder could possibly enter undetected.

Slow-moving targets can be a problem. A PIR sensor will not detect very slow motion. However, defeating a PIR sensor with slow motion is difficult to do because a person must keep all body movement to a minimum.

A PIR sensor may fail to detect movement in an area if the lens is masked or fogged.

4.3.5 Installation Criteria

Installation of PIR sensors is fairly inexpensive. The manufacturer's guidelines should be followed as appropriate.

All hot spots that generate infrared energy should be removed or shielded. Radiant energy from such sources may produce thermal gradients that will change the background energy pattern. Hot spots can be open heating elements, incandescent light bulbs, convective heat currents, and direct sunlight on windows, floors, and walls. Sunlight can enter the protected area directly through openings, such as broken window panes, ventilation grids, and poorly fitting doors.

For optimal intruder detection, the sensor should be aimed so that the path likely taken will be across the sensor field of view, rather than toward or away from the sensor.

To prevent an intruder from circumventing the sensor, its detection envelope should not be smaller than the physical boundaries of the area being protected. The detector should not be mounted directly above a doorway or a window or in any position that would allow an intruder access to the sensor from beneath the sensor.

For high-security applications, the small LED light on the sensor that indicates a detection should be turned off when not being tested by authorized personnel.

Placing the sensor near a light source can generate nuisance alarms caused by insects attracted to the light.

All sensors should be provided with the following:

- Supervised wiring in conduit

- Fail-safe operation

- Emergency power in case of main power failure

- Tamper indication

An end-to-end self-test is desirable. A final test should be performed after installation to verify the sensor coverage area.

4.3.6 Testing

A regular program of testing sensors is imperative for maintaining them in optimal operating order. Three types of testing need to be performed at different times in the life of a sensor: acceptance testing, performance testing, and operability testing.

4.3.6.1 Acceptance Testing

When a PIR sensor is first installed, it should be tested in order to formally "accept" the sensor as part of the physical protection system. Acceptance testing consists of two parts:

(1) A **physical inspection** to ensure that the sensor was installed properly consists of the following:

- Verify that the installation matches the installation drawings, which should follow the guidance provided by the manufacturer.
- Verify that signal and power wires are routed in the conduit.
- Verify proper power levels (voltage and amperage).
- Verify correct wire connections.

(2) A **performance test** to establish and document the level of performance consists of following the performance testing procedure (described below for the recommended tests.

4.3.6.2 Performance Testing

Performance tests are intended to verify that the level of performance of each PIR sensor is consistent with the documented performance achieved during the original acceptance testing.

Performance testing should be conducted whenever an electronics module is replaced, the optical alignment is changed, or an adjustment is made that can affect sensitivity. This test should include a visual inspection of the sensor and of the general area where the sensor is installed. Personnel conducting the test should refer to the maintenance section and perform the prudent routine maintenance. Test procedures recommended by the manufacturer should be followed. As in all test situations, the area being tested should either be kept under visual observation by a member of the site security force, or a member of the site security force should conduct the test.

For each area of detection, the test should do the following:

- Ensure that the system meets the manufacturer's specifications and recommended detection probability.

- Verify that no dead spots exist in the zone of protection.

- Verify that line supervision and tamper-indication alarms in both the access and secure modes are functional.

- Verify that both line supervision and tamper-indication alarms are received in the alarm station as appropriate.

Records of initial testing capabilities, equipment sensitivity settings, or voltage outputs should be maintained so that deterioration in equipment capability can be identified and monitored.

Walk tests should be performed for all areas covered by the PIR sensor and compared with the results of the initial acceptance test to check for any degradation in the coverage of the sensor.

4.3.6.2.1 Tangential (Arc) Path Testing

The following instructions describe the walk test to be conducted along tangential paths (refer to Figure 48); this approach has the likeliest chance of detection as the PIR sensor is most sensitive in this direction.

(1) Start outside of the published detection area on one side and walk at 1 foot per second along the first tangential path.

(2) Stop when an alarm occurs and mark that position.

(3) Return to the starting point, wait 30 seconds for the sensor to reset, and repeat walk test along the same path.

(4) Repeat testing on that path until the required number of tests is completed. Multiple tests along each test line path are required to establish a P_D (probability of detection). For example, in order to establish that a sensor has a minimum P_D of 90 percent at a confidence level of 95 percent, the sensor would have to pass 29 out of 30 tests.

(5) Perform the steps above on remaining paths.

(6) Repeat Steps 1 through 5, starting from the other side of the detection area.

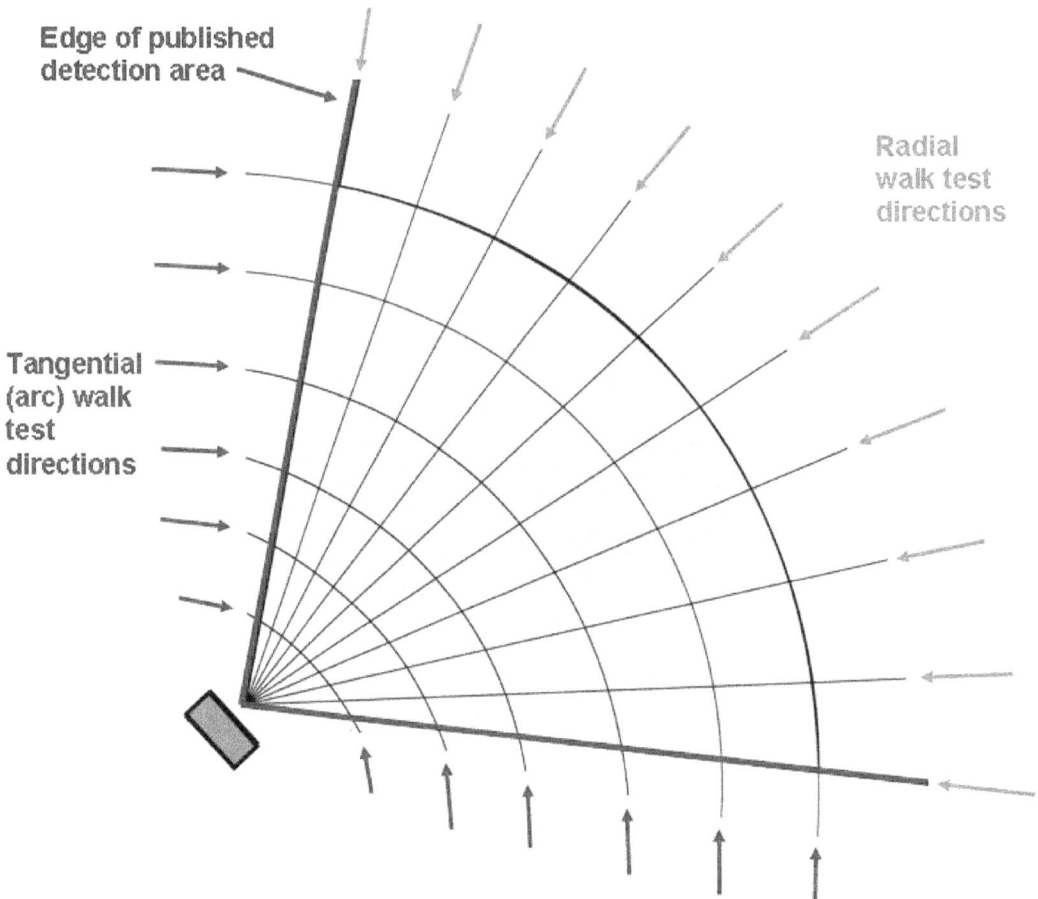

Figure 48: Recommended walk test paths and directions for the performance testing of a PIR sensor.

4.3.6.2.2 Radial Path Testing

The following instructions describe the walk test to be conducted along the radial paths (refer to Figure 49); this approach has the least likely chance of detection as the PIR sensor is least sensitive in this direction.

(1) Start outside of the published detection area in front of the sensor and walk at 1 foot per second along the first radial path.

(2) Stop when an alarm occurs and mark that position.

(3) Return to the start point, wait 30 seconds for the sensor to reset, and repeat walk test along the same path.

(4) Repeat testing on that path until the required number of tests is completed. Multiple tests along each test line path are required to establish a P_D (probability of detection). For example, in order to establish that a sensor has a minimum P_D of 90 percent at a confidence level of 95 percent, the sensor would have to pass 29 out of 30 tests.

(5) Perform Steps 1 through 4 for the remaining radial paths.

Changes in room configuration can affect sensor coverage and should be checked. If room configuration has changed significantly, a complete retest of the sensor coverage should be initiated to ensure the protection of the room or the asset.

The test should determine the most vulnerable area for each section and the method of approach most likely to penetrate (e.g., walking, running, jumping, crawling, rolling, or climbing). This determination will, in most cases, be sensor and location dependent. The penetration approach that is most difficult to detect should be attempted more frequently if an equal number of tests for each approach is not possible.

The various approach paths should be tested in random order, which will preclude the possibility that environmental effects and other unknown factors are affecting the test results (detection or nondetection). Use of a random sequence reduces the chance that the test results will be biased.

When a designer is determining the best place to locate an asset within a room, the enclosed sample graph showing the PIR sensor detection pattern could be used as guidance (refer to Figure 49). The asset should be placed inside the area that has a high PD, no matter the direction from which the intruder approaches the sensor.

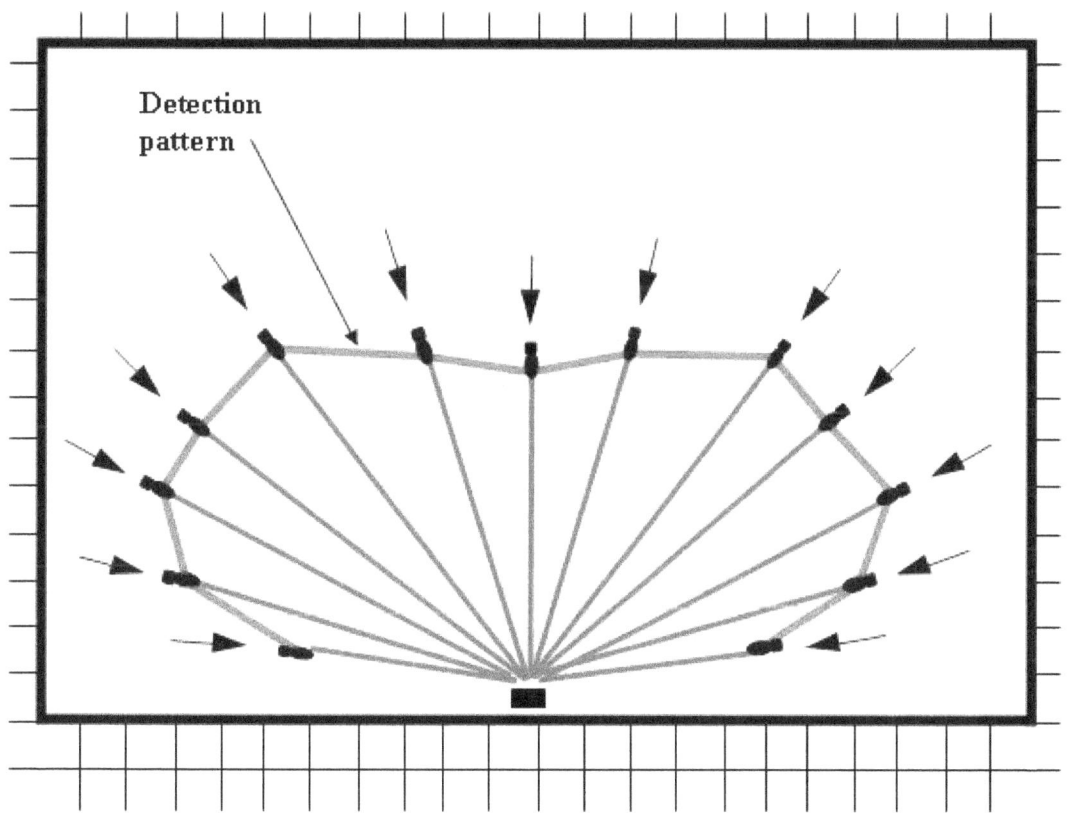

Figure 49: A typical PIR detection pattern derived from walk tests towards the sensor.

4.3.6.3 Operability Testing

Operability testing should be conducted by crossing the zone of detection in the area that the sensor is installed to monitor. The detection capabilities of each PIR should be walk tested in a different, preferably random, order every 7 days, and the tests should be conducted throughout the week rather than all on the same day. The testing should result in 100-percent detection on all segments every 7 days. If the interior intrusion alarm system fails to detect an intrusion in one or more segments, corrective actions should be taken and documented. Records should be maintained to document that all required testing has been accomplished.

4.3.7 Maintenance

The general maintenance guidelines outlined in the manufacturer's technical manuals should be followed on a schedule determined by security maintenance staff and security forces, as well as by the environment in which the sensor is installed.

At a minimum, the sensor optics should be periodically cleaned and frequent visual inspections performed to ensure that no objects have been moved into a blocking position that would render the sensor inoperative.

Equipment maintenance guidelines generally recommend keeping 10 to 20 percent of spare parts on hand, based on total facility units. This requirement may be adjusted as maintenance data are accumulated on the failure rate of specific sensors and sensor components. If replacement parts can be obtained quickly from regional distributors, smaller onsite inventories would be adequate.

4.4 Proximity Sensors

4.4.1 Principles of Operation

Proximity sensors, also known as point protection/detection devices, have the capability of detecting someone approaching, touching, or attempting to remove valuable items. Proximity sensors usually form the innermost level of protection, after exterior perimeter sensors, boundary penetration sensors, and/or volumetric sensors. Since they are usually located close to a particular asset, the response force has the least amount of time to respond to an alarm once the intruder is detected. Because of this, proximity sensors should not be used as primary detection on a high-risk item. Proximity sensors are most effective for protection against an insider.

4.4.2 Types of Proximity Sensors

Types of proximity sensors include the following:

- Capacitance

- Pressure

- Strain

- Switches

The various types of proximity sensors are described below.

4.4.2.1   Capacitance

Capacitance proximity sensors operate on the same principle as an electrical capacitor. These types of detectors are used to protect metal containers that can be isolated from ground such as safes or file cabinets. An electrical capacitor comprises one or more conductors separated by a dielectric medium. A change in the electrical characteristics of the dielectric medium causes a change in the capacitance between the two plates. In the case of the capacitance proximity sensor, the protected metal object corresponds to one plate, and an electrical reference ground plane under or around the protected object corresponds to the second plate. An insulator isolates the protected object from ground. The air between the object and ground comprises the dielectric medium. When a person comes close to, or touches the object, the dielectric is changed, which changes the capacitance. The processor (part of the capacitance sensor) detects the change in capacitance and generates an alarm.

4.4.2.2   Pressure

Pressure sensors incorporate a sensing device that responds to deformation of the sensor caused by weight placed on it. Pressure mats consist of a series of ribbon switches positioned parallel to each other, approximately 3 inches apart along the length of the mat. Ribbon switches are constructed from two strips of metal in the form of a ribbon separated by an insulating material. They are constructed so that when an adequate amount of pressure, depending on the application, is exerted anywhere along the ribbon, the metal strips make electrical contact. They can be used to detect the presence of intruders when they approach or attempt to move protected items. For instance, pressure mats can be installed under the carpet around the protected item. Then anyone who approaches the item steps on the mat and initiates an alarm. The operation of a pressure mat represents the operation of pressure sensors in general. The pressure sensor output signal is routed to an alarm console to indicate an intrusion.

4.4.2.3   Strain

Strain sensors measure small amounts of deformation or flexing of a surface. A basic configuration of a strain sensor would be one or more sensing devices connected to a processor. The sensing devices can employ piezoelectric, piezoresistive, metal foil, or wire to detect surface deformation or flexing. When configured to be a strain sensor, the electrical properties (such as resistance) of these materials will change when bent, stretched, or compressed. The processor continually measures the electrical properties of the sensing device and will output an alarm or other indication when a specified amount of change has occurred. Processors will typically have user-programmable controls to specify how much change has to occur in order to output an alarm. In a proximity sensor application, the sensing device is attached to a surface that is flexed slightly when an object (such as a protected item) is placed on it. The processor is programmed to alarm if there is a change, such as when the object is removed or tampered with.

4.4.2.4   Switches

Switches can be used as a proximity point sensor. A protected item is placed on the switch, actuating it so that the electrical contacts are either in an open or closed position. Alarm system electronics monitors the switch for a change in the position of the contacts. If the item is removed, the contacts change position and an alarm is generated. Movement of the item can also cause an alarm if the movement is such that it causes the switch contacts to change positions. The surface and switch mounting need to be designed so that it is very difficult to remove the protected item while maintaining the switch in the secured position.

4.4.3 Sources of Nuisance Alarms

The sensitivity of capacitance sensors is affected by changes in relative humidity and the relocation of other metal objects closer to or farther away from the protected item. Changes in the relative humidity vary the dielectric characteristics. A rapid increase in humidity causes the dielectric (air) conductivity to increase and reduces the capacitance, resulting in an alarm. Conversely, a decrease in humidity or drying of the air reduces the conductivity. Similarly, when larger metal objects (high electrical conductivity) such as cabinets, desks, equipment racks, etc. are moved close to an object protected by a capacitive sensor, the sensitivity of the sensor can change. If the sensitivity is increased, the chance for nuisance alarms increases. If the sensitivity is lowered, detection capability is lowered.

Nuisance alarms from pressure mat sensors can occur if the insulating or separating material that keeps the ribbon switch contacts apart deteriorates because it is worn out or exposed to harsh conditions. Extreme heating and cooling (out of the operating range) are additional nuisance alarm sources, especially if the mat is worn and deteriorated. A mat installed near heavy traffic areas where workers or other personnel would inadvertently step on it is an additional source of nuisance alarms. Pressure mats that are in good condition and installed properly should have very few nuisance alarms.

Strain sensor devices can be affected by changes in temperature, which result in nuisance alarms. Some strain sensors are configured to reduce or eliminate the effects of temperature changes. Changes in humidity can also be a source of nuisance alarms. If the protected item can absorb moisture, the weight of that object could change enough with humidity changes to cause nuisance alarms.

Switch sensors in good condition and installed properly should have very few, if any, nuisance alarms. Primary nuisance sources include loose or damaged mounting brackets, fasteners, and mounts and damaged or worn out internal or external components of the switch itself.

4.4.4 Characteristics and Applications

Proximity sensors should not be used as primary detection for high-risk items. They are typically used as a second or third line (layer) of protection and are most effective for protection against an insider. The goal would be to detect the insider who is very close to, touching, or moving an object that he or she should not have access to. Portable high-value objects should be in a cage or safe or tied down to add delay to an abrupt theft. (See Figure 50.) Portable, high-risk/value items should be locked in cabinets or secured by other means to add delay in removal. Proximity sensors can be installed to detect attempted removal of tiedowns or opening of cabinets.

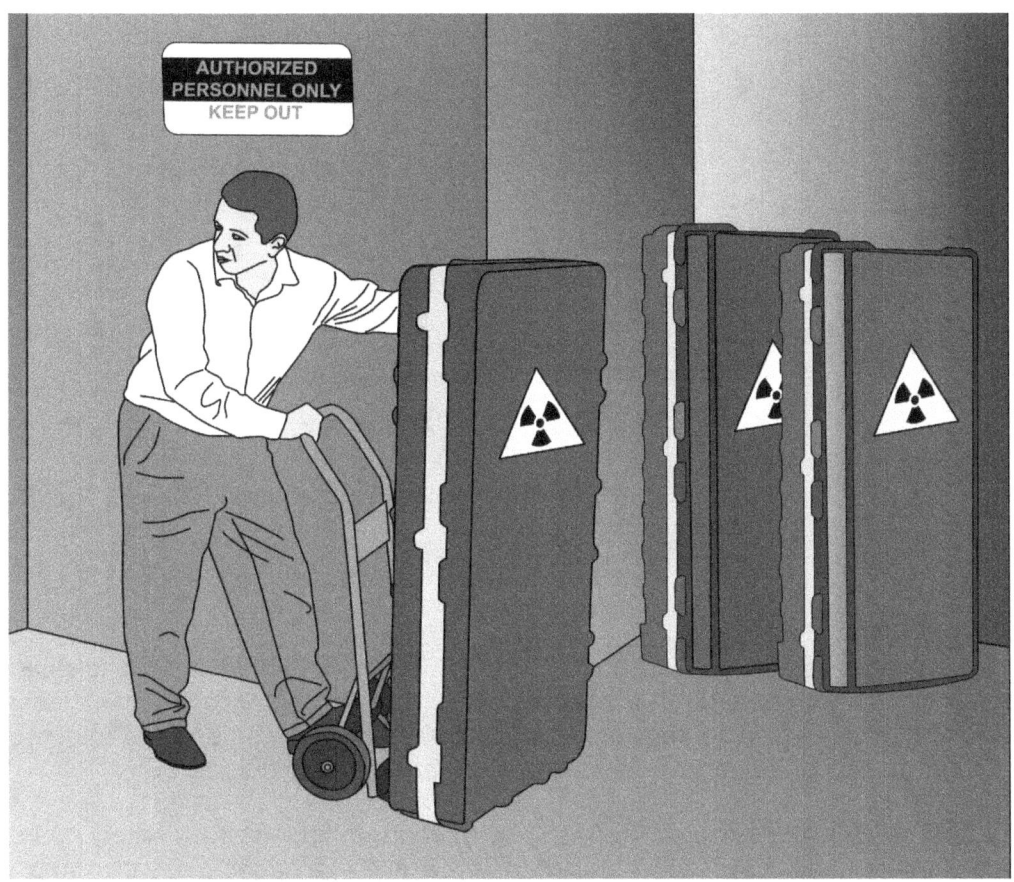

Figure 50: Example of portable, high-value items.

A typical application of a capacitance proximity detector would be the protection of a safe or file cabinets (see Figure 51). The safe or file cabinets must be set on blocks to isolate them from the ground plane. The blocks should be made of a nonconductive plastic or nonhydroscopic material. Wooden blocks should not be used because they are hydroscopic and could absorb enough moisture over a period of time to change the dielectric enough that the protected objects become insensitive. Capacitive detectors can be used to protect paintings, tapestries, and other objects by installing a relatively large copper foil sheet or metal screen under the objects requiring protection. In this type of application, the metal screen becomes part of the protected circuit, as is the safe or any other metal object.

Pressure mats are commonly used in industrial and commercial applications such as controls for opening doors or as safety devices for machinery. Although less common in security applications, they can be used along probable intruder routes or around valuable objects. They are usually well concealed under carpets or flooring to make it more difficult for an intruder to determine the location of the detection area.

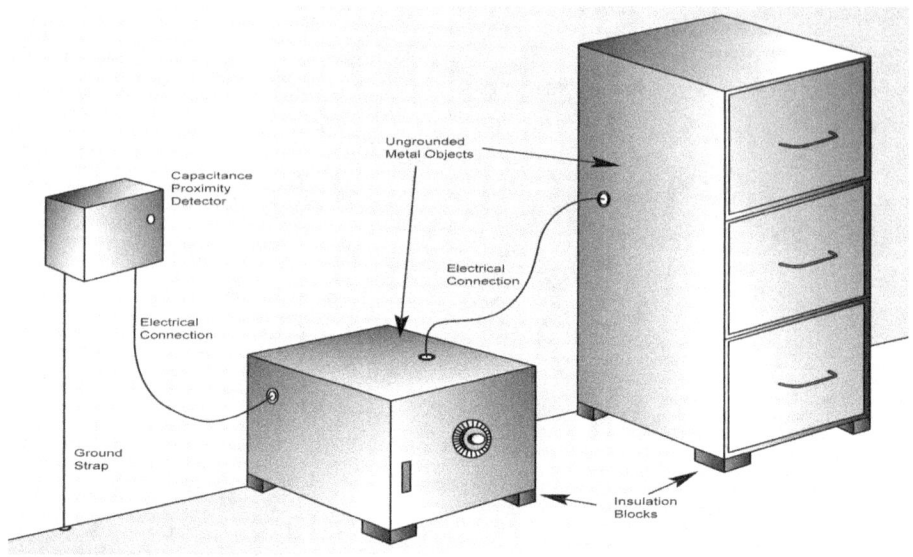

Figure 51: Example of a capacitance sensor installation.

Strain sensors can be used to continually monitor the weight of an object, sensing when it is being lifted, moved, or tampered with (Figure 52). The environment, such as changes in temperature or humidity, must be considered. Strain sensors can also be used to detect a person's weight as he or she approaches a protected area or item (Figure 53).

A self-adjusting capacitance proximity sensor may be defeated by extremely slow approach to the protected object. An adversary can detect the presence of a capacitance proximity sensor by employing radiofrequency field-sensing equipment, such as that used by a telephone company to find underground telephone cables. Relocating any conducting object closer to or farther away from the protected surface can cause a small capacitance change between the protected surface and ground, generating a nuisance alarm. These objects include persons walking near or leaning on the surface, cabinets or other objects being moved close to the surface, or loose-fitting components of the protected object itself.

Figure 52: Pressure mat example.

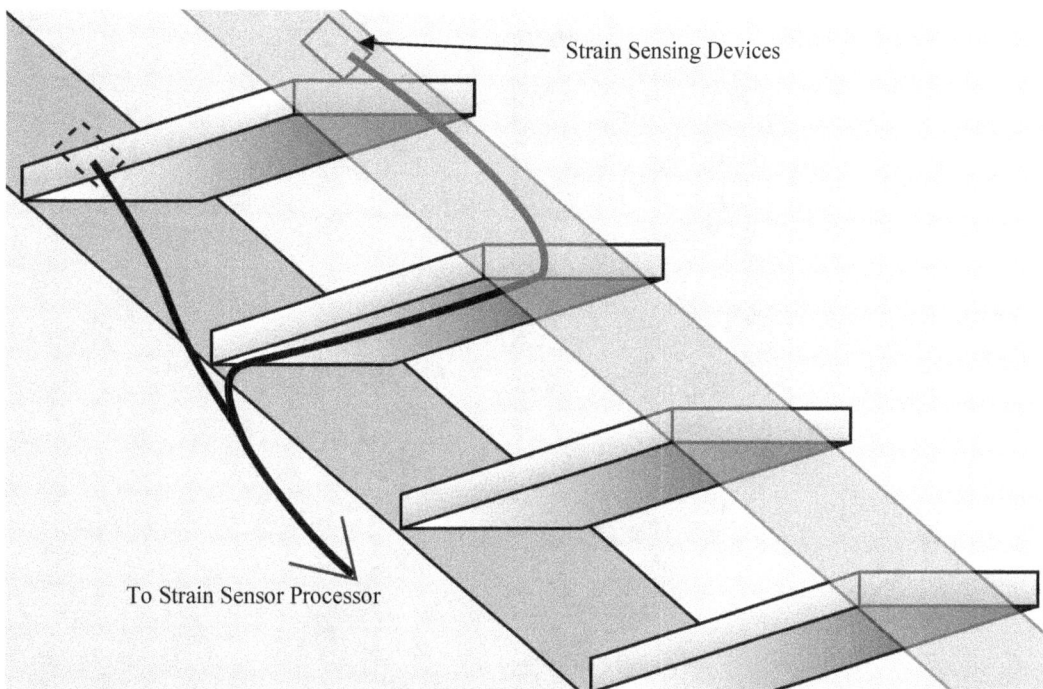

Figure 53: Strain sensor concept to detect person on stairs.

A capacitance sensor may not work well (i.e., high nuisance alarms) if the protected object is in an area where there is high traffic close to the object when the sensor is in the secured condition.

The sensitivity of a capacitance proximity sensor is affected by sudden changes in the relative humidity. Changes in the moisture content of the air will vary the dielectric characteristics by either increasing or decreasing its conductivity. If the sensor sensitivity is adjusted to detect an intruder several meters from the object, the change in conductivity may be enough to initiate a nuisance alarm. Capacitance proximity sensors employing a self-balancing circuit adjust automatically to changes in relative humidity and to relocation of conducting objects near the protected object.

A pressure sensor is vulnerable to bridging by a board placed on bricks or by jumping or stepping across it. A pressure sensor also is subject to considerable wear from normal traffic, and periodic tests should be performed to ensure that the sensor is operating effectively. The example mat sensor shown in Figure 54 could be used to let room occupants know that someone is at the door, but it can be easily bridged if installed this way for security applications.

A strain sensor responds to any action that causes the surface upon which it is mounted to flex. Heavy machinery in the building or nearby heavy vehicular traffic can cause surfaces to vibrate, which may result in nuisance alarms. Although the sensor operates at frequencies as low as direct current, limiting the low-frequency response avoids alarms caused by long-term drift and slow deformation of the structure over time.

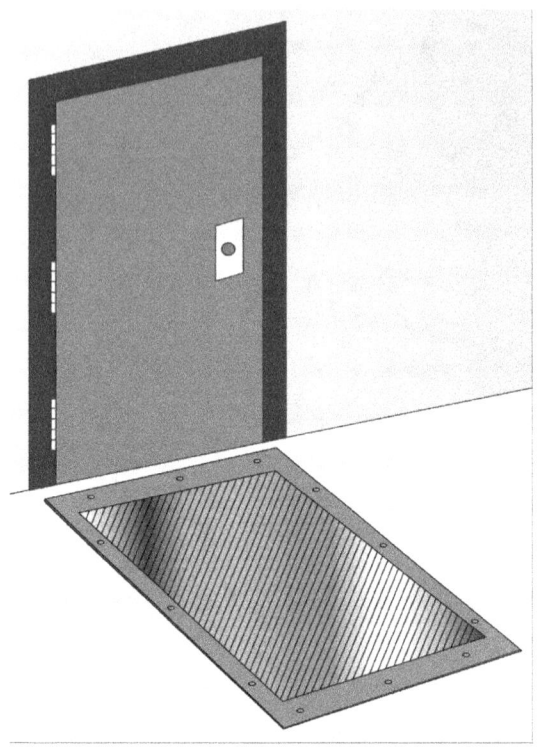

Figure 54: Mat sensor near door can be bridged if it is not installed correctly.

4.4.5 General Installation Criteria

For installation and setup of proximity sensors, manufacturer instructions and recommendations should be followed. The sensors should be equipped with a tamper-indicating device that is continuously monitored by the security system. These tamper-indicating devices are typically switches that detect a cover or door (that allows access to the sensor electronics) being removed or opened. All wiring for these systems (alarm, power, or tamper) should be in conduit, and the alarm and tamper signal wiring should be supervised. Backup batteries or standby power is recommended. Once installed, performance testing is necessary to confirm desired detection capability. Testing should include attempting to defeat the sensor using defeat methods for the sensor technology. Short-term trial operation (several days to several weeks) will help determine if there are any initial and common nuisance alarm sources. An end-to-end self test is also desirable. General installation criteria for each type of proximity sensor follow.

4.4.5.1 Capacitance Sensor

The capacitive sensor will need a ground plane, which could consist of cables, conductive mats, or conductive foil under, near, or around the protected object. The protected object will need to be isolated from the ground plane using nonhydroscopic materials. Both the protected object(s) and the ground plane will need to be connected to the processor unit, which is typically mounted on a wall close to the protected object.

4.4.5.2 Pressure Sensor

The pressure sensor is usually concealed from an intruder as a doormat or by placing it under the floor covering. The sensor pad can be installed in a depression in the floor. If the pad is placed under a protective cover, such as a rug or a rubber doormat, the protective cover must be fastened down around the edges to prevent the pad from moving or being removed.

4.4.5.3 Strain Sensor

Generally the sensing device is mounted at the point where the largest deflection is most likely to occur. If that point cannot be defined, the best procedure is to mount the sensor in the center of the surface. The sensor is bonded to the surface as rigidly as possible, so when the surface flexes, the sensor will be forced to elongate or contract and will not separate or slide along the surface. Site- or object-specific design and fabrication of the mounting or attaching surface may be required.

4.4.5.4 Switch Sensor

This sensor may also require site- or object-specific design and fabrication. Basic installation criteria include a way to secure the object to the placement surface, protection of the switch and switch wiring from tampering, protection to make it very difficult to remove or move the object while maintaining the switch in the secured position, and connection of the switch contacts to an alarm communication system.

4.4.6 Sensor Testing

4.4.6.1 Acceptance Testing

When a sensor is first installed, it should be tested in order to formally "accept" the sensor as part of the physical protection system. Acceptance testing consists of two parts:

(1) A **physical inspection** to ensure that the sensor is installed properly consists of the following:

  - Verify that the installation matches the site installation documentation/ drawings, which should follow the guidance provided by the manufacturer.
  - Verify that signal and power wires are routed in the conduit.
  - Verify proper power levels (voltage and amperage).
  - Verify correct wire connections.

(2) **Performance testing** to establish and document the level of performance is described below.

4.4.6.2 Performance Testing

Performance testing should include a visual inspection of the sensor and of the general area where the sensor is installed. When performing testing, alignment, or adjustments on sensors, the assistance of additional personnel should be considered as these activities can be difficult for a single person to manage.

For proximity sensors, performance testing should include tests to verify good detection when a protected item and any associated delay hardware is approached, touched, or moved. Specific tests will depend on the sensor type and installation configuration. Tests should be conducted at different locations around, near, and at the protected object. Several tests at each location should be conducted to achieve a confidence level for detection. Performance testing procedures should include testing of tamper switches, backup batteries, or power supplies and receipt of correct detection and tamper alarms at the alarm stations.

Sensor-specific performance test procedures should be developed and documented. Procedures should include step-by-step test methods, pages or forms to record test results (including a sketch of the sensor detection coverage if applicable), a form to record the sensor model, serial number, settings, and other data as needed. Test results and data can be compared to previous test results to determine trends or the occurrence of gradual degradation. Test results and documentation should be protected from unauthorized disclosure.

4.4.6.3 Operability Testing

Operability tests should be conducted on proximity devices in a manner in which the sensor is designed and installed to function (e.g., activating sensor alarms by opening doors, moving through sensored areas, or being close to or touching an object such as a safe). Certain proximity sensors require a protected item to be moved or a protective cage or cabinet to be opened to activate the alarm (e.g., switch and strain sensors). Testing of these types of alarms may require increased coordination with facility personnel to ensure that safety and security are maintained during the testing.

During operability testing, the protected area or asset can be visually inspected to ensure that objects have not been placed in the area or near the protected asset that affect the detection capabilities of the sensor or could cause nuisance alarms.

4.4.7 Maintenance

Periodic operability and performance tests, in addition to any self-test invoked by the sensor or the system, need to be performed to ensure that the sensor is operating effectively. A visual inspection of the installation should be performed periodically, particularly after any major maintenance to the protected surfaces. Standby batteries should be tested on a conservative schedule and replaced when indicated. Every maintenance or repair action should be entered in a log to record the date, time, corrective action, who performed the maintenance or repair, and an assessment of what may have caused the problem. Maintenance activities for each sensor are described below.

4.4.7.1 Capacitance Sensor

Extremely good housekeeping is required in the area near the protected object because a capacitance proximity sensor is very sensitive to the environment within a few inches of the protected surface. Any conducting object large enough to change the dielectric of the air that is placed near the protected item must be removed. Wet mopping or liquids spilled on wooden floors under and around the protected object can significantly change the operation of a capacitance sensor.

4.4.7.2 *Pressure Sensor*

A mat-type sensor can be subject to considerable wear from traffic during normal business hours, so frequent operability tests are important. During testing, the sensor should be inspected for visible signs of wear and degradation.

4.4.7.3 *Strain Sensors and Switch Sensors*

During periodic testing, these sensors should be inspected for loose attaching hardware, loose connections and excessively worn parts.

4.5   Dual-Technology Sensors

4.5.1 Principles of Operation

Dual-technology sensors (also referred to as "dual techs") were designed to lower the false or nuisance alarm rates in an interior sensor. This is accomplished by combining two different types of sensors in one casing so that each sensor is complementary: each sensor generates a different set of nuisance alarm sources. The two sensors are connected electronically by using an "AND" gate logic function; both technologies need to sense an event within a predetermined interval before a valid alarm will be generated. If one technology has a detection but the other does not, no alarm will be generated. Because the two sensors will not sense an intrusion at the same instant, the system is designed to generate an alarm when both sensors sense an intrusion in a preselected time interval, usually a few seconds. This time interval is usually a parameter that the user can configure.

Reducing the nuisance alarm rate of a sensor is highly desirable. However, this feature comes at a price: making a unit less sensitive to possible nuisance alarms also makes the unit less sensitive to valid alarm conditions. Because of this possibility, dual-technology sensors configured to operate in an "AND" gate logic are not normally recommended for high-risk or high-security facilities.

Less commonly, dual-technology sensors can be designed to operate using "OR" gate logic. With the OR configuration, either sensor technology can generate an alarm independent of the other. This configuration is similar to having two separate sensors installed in the same location. Unfortunately, two different types of sensors are not likely to both be optimally installed in the same location. For example, a PIR sensor would be best placed such that an adversary would likely walk across the detection zone, while a microwave sensor would be best positioned such that an adversary would likely walk toward or away from the detection unit. Therefore, in high-risk facilities, the installation of two different types of sensors in locations that are optimal to their own detection capabilities would be preferable to the use of a dual-technology sensor configured with an "OR" gate logic. (See Figure 55.)

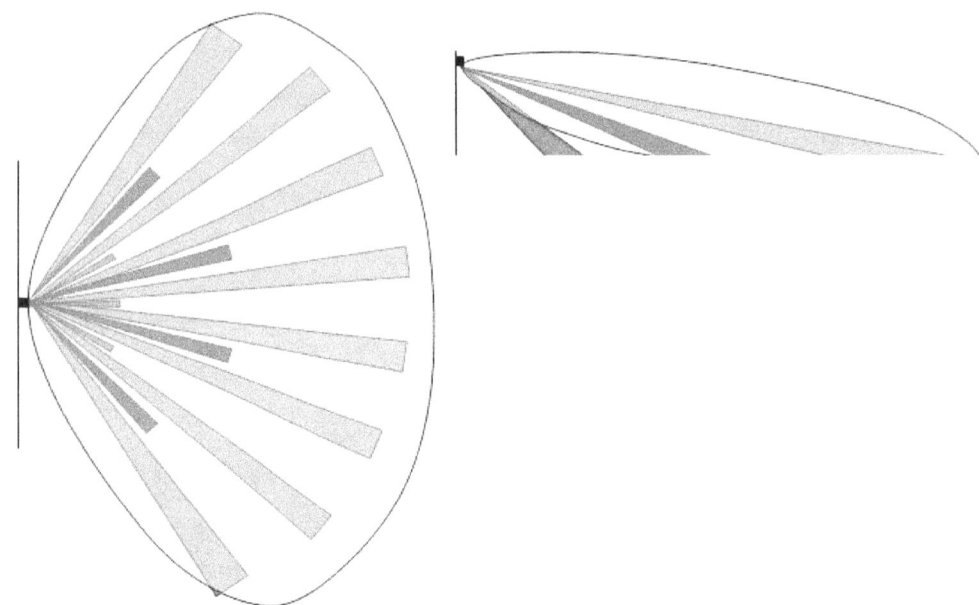

Figure 55: The figure on the left depicts an overhead view, and the figure on the right depicts a side view, of the patterns of detection for a dual-technology sensor comprising a PIR sensor and a microwave sensor; the pale yellow "bubble" illustrates the microwave detection pattern, and the remaining "fingers" illustrate the PIR detection pattern. Note that most dual-technology sensors use "AND" logic for the two different sensors, such that both sensors would have to sense a detection before the sensor would signal an actual alarm.

4.5.2 Types of Dual-Technology Sensors

Two combinations of dual-tech sensors are usually used for interior protection: Passive Infrared Acoustic and Passive Infrared Microwave. Numerous manufacturers build these sensors with varying technical specifications so care should be taken in selecting the sensor that meets the needs of the particular application.

4.5.2.1   Passive Infrared Acoustic

The PIR-acoustic dual-technology unit employs a PIR sensor and usually an ultrasonic sensor. The ultrasonic sensor generates a teardrop-shaped pattern of acoustical energy at frequencies well above those audible to humans. The patterns are typically about 9 meters (30 feet) deep and 7.6 meters (25 feet) wide. They detect disturbances in the reflected energy from anyone moving in a radial direction in the energy pattern. The shape of the PIR detection segments is similar to the ultrasonic energy pattern configuration. Someone moving in the zone of detection will create an alarm if detected by both the PIR sensor and ultrasonic sensor.

4.5.2.2   Passive Infrared Microwave

The PIR-microwave dual-technology unit employs a PIR sensor and a microwave sensor. The microwave sensor detects Doppler changes in the reflected microwave energy pattern produced by someone moving in the area with some radial velocity relative to the detector. This teardrop-shaped pattern typically covers an area about 20 meters (65 feet) deep and 15 meters (50 feet) wide. The shape of the PIR detection segments is also similar to the microwave energy pattern configuration (Figure 55).

4.5.3 Sources of Nuisance Alarms

Because dual-technology sensors are normally configured to operate with "AND" logic, the likelihood of nuisance alarms is greatly reduced. If operated with "OR" logic, sources of nuisance alarms will be the sum of sources of nuisance alarms for the individual sensors.

The PIR detector within most dual-technology units can trigger false alarms from sources of heat, sunshine, and incandescent lights, as well as other sources. (See "Sources of Nuisance Alarms in Passive Infrared Sensors" in Section 4.3 of this report.)

For the PIR-acoustic dual-tech sensor, the ultrasonic sensor effectiveness is reduced by air turbulence from heating or air-conditioning ducts, drafts, or other sources of moving air. Acoustic energy generated by ringing bells and hissing noises, such as the noises produced by radiators or compressed air systems, contains frequency components in the operating frequency band of ultrasonic motion detectors. These sources of ultrasonic energy may occasionally produce signals similar to an intruder, which can confuse the signal processor and result in nuisance alarms.

For a PIR-microwave dual-tech sensor, one important characteristic to remember is that microwave energy can pass through light construction material, which can be a source of nuisance alarms.

It is also important to remember that for a dual-technology sensor, when two sensors are logically combined using an "AND" gate, the probability of detection of the combined detectors will be less than the probability of detection of the individual detectors. The probability of detection of the combined sensors in a single unit will be less than if the individual detectors are mounted perpendicular to each other with overlapping energy patterns and fields of view.

4.5.4 Characteristics and Applications

Dual-tech sensors usually have a lower nuisance alarm rate than single technology sensors when the detectors are properly applied and assuming that each has a low nuisance alarm rate. This sensor type attempts to achieve absolute alarm confirmation (i.e., no nuisance alarms) while maintaining the highest probability of detection (P_D) possible for this kind of unit.

The main advantage of a PIR-microwave dual-tech sensor is that both the PIR and the microwave are complementary in providing long, narrow fields of detection. The false alarm rate is reduced significantly by the combination of the technologies in the AND configuration. These types of sensors would best be used as a proximity-type sensor where detection is confined to a small area or a single object.

A PIR-acoustic dual-tech sensor can typically cover open areas approximately 7.6 meters (25 feet) by 7.6 meters (25 feet). A feature of ultrasonic energy is that it will not penetrate physical barriers such as walls; therefore, it can be easily contained in closed rooms. Since acoustical energy will not penetrate physical barriers, the walls of the protected room either absorb or reflect the energy.

Ceiling-mounted transceivers generate a cone-shaped energy pattern that can cover a circular area about 9.1 meters (30 feet) in diameter when the transceiver is mounted 3 to 4.6 meters (10 to 15 feet) above the floor. A ceiling-mounted transceiver can be mounted directly over the area requiring protection. This feature is especially valuable in areas where it is difficult to protect using wall-mounted transceivers. Long-range ultrasonic transceivers are available with long, narrow energy patterns for protecting aisles and hallways. A single detector of this type can protect a hallway about 21.3 meters (70 feet) long. Combined microwave and infrared detectors cover open areas from 12.2 meters (40 feet) deep and 7.6 meters (25 feet) wide up to 22.9 meters (75 feet) deep and 13.7 meters (45 feet) wide as well as long, narrow areas up to 61 meters (200 feet) long.

When sensors are combined in a logical "AND" configuration, the P_D of the combined detectors is less than the P_D of the individual detectors. If an ultrasonic sensor with a 0.95 P_D is combined with a PIR sensor having a 0.95 P_D, then the resulting 0.90 P_D for the dual–tech sensor is the product of the individual probabilities. A P_D of 0.90 may not meet the required probability of detection for some facilities.

Assuming a single direction of intrusion, a higher P_D can be obtained from separately mounted sensors than from a dual-technology sensor. Ultrasonic and microwave sensors have their highest P_D for radial motion either toward or away from the sensor, but a PIR sensor has its highest P_D for motion circumferentially across its field of view. Thus, the P_D for the sensors combined in a single unit and aimed in the same direction is less than the P_D for individual detectors mounted perpendicular to each other with overlapping detection envelopes. The highest P_D for a dual-tech sensor is achieved by treating the individual sensors separately, in an "OR" configuration.

Vulnerabilities for a dual-technology sensor include vulnerabilities for each sensor technology. In the "AND" configuration, if either sensor is defeated, then the dual-technology sensor unit is defeated. For this reason, a dual-technology sensor should never be used as a replacement for two separately installed sensors in high-security applications. If a dual-technology sensor is necessary in a location because of nuisance alarm issues, then another sensor (dual- or single-technology) should be used and installed in such a manner that each sensor unit protects the other, as well as providing overlapping detection coverage within the area being protected.

It is important to avoid environments where either detector type would be prone to false alarms. If either detector is exposed to an environment where it experiences a high number of false alarms, then the probability of one of these false alarms being present at the logic AND gate when a false alarm arrives from the second, and perhaps more stable, sensor is high. For example, if a PIR-acoustic detector was installed in a drafty area where the ultrasonic detector would experience a high number of false alarms because of the distortions in its projected energy pattern and the infrared detector might experience a few alarms as the result of background temperature changes caused by the drafts, the probability of simultaneous alarms from both sensors would be increased. A combination microwave and infrared detector would be a better choice for such an application because the microwave detector would not be affected by this environment. The drafts would still cause temperature changes that could affect the infrared detector, but since it would be combined with the microwave detector, the probability of simultaneous false alarms would be low and, consequently, the false alarm rate would be lower.

One concern when using a dual-tech sensor that combines microwave and infrared detectors with "AND" logic is that the microwave's detection zone is usually much larger than the infrared's detection zone; hence, no detection will occur until the adversary reaches the point where both sensors can detect.

4.5.5 Installation Criteria

Microwave technology is usually more sensitive in its least sensitive direction than the PIR in its least sensitive direction. Because of this, the following considerations apply:

- Performance testing and evaluation should be similar to that of a PIR sensor.

- The sensor should be installed with primary consideration given to the PIR section. The most likely paths to the protected item or area should cross through the PIR detection pattern and not be directly toward or away from the sensor unit.

- A dual-technology sensor should be located so that the likely path of an adversary will be across the sensor detection area and less likely to be toward the sensor.

A dual-technology sensor should be installed using a sturdy mount. Vibration can cause misalignment or make the sensor prone to nuisance alarms. The installer should make sure that the sensor is aligned away from possible nuisance alarm sources such as heaters.

If dual-technology sensors are to be used, multiple sensor units should be installed, with each unit offering overlap protection of the other.

4.5.6 Testing

A regular program of testing sensors is imperative for maintaining them in optimal operating condition. Three types of testing need to be performed at different times in the life of a sensor: acceptance testing, performance testing, and operability testing.

4.5.6.1 Acceptance Testing

When a dual-technology sensor is first installed, it should be tested in order to formally "accept" the sensor as part of the physical protection system. Acceptance testing consists of two parts:

(1) A **physical inspection** to ensure that the sensor is installed properly, consisting of the following:

- Verify that the installation matches the installation drawings (and the drawings should follow the installation guidance provided by the manufacturer).
- Verify sensor intersection spacing.
- Verify that signal and power wires are routed in conduit.
- Verify proper power (voltage, amperage) levels.
- Verify correct wire connections.
- Perform a complete alignment in accordance with the manufacturer's manual on all units to verify that all modules are operational and oriented correctly. The method of physical alignment is specific for each manufacturer's model.

(2) **Performance testing** should establish and document the baseline level of performance (see below for a description of the recommended tests).

4.5.6.2   Performance Testing

Performance testing and evaluation should be similar to that of a PIR sensor. (See Section 4.3.6.2.) The most likely path to the protected item or area should ideally cross through the PIR detection pattern and not be directly towards or away from the sensor unit. In addition to the extensive walk tests described in the PIR section of this report, additional tests can be performed. Slow walk tests are conducted at speeds less than 0.5 feet per second. Most volumetric sensors will have a speed at which detection capability decreases. If the potential to circumvent a system by crawling is a concern, crawl testing should be performed to obtain detection characteristics. The detection pattern of a crawling person will likely be different than that of a walking person.

Performance tests are designed to verify the level of performance of each dual-technology sensor through the range of its intended function. Performance testing should be conducted when an electronics module is replaced, when there is a change of the physical alignment or any adjustment that can affect sensitivity, after remodeling of the building structure, or any major change in the arrangement of furniture or equipment. The performance test should include a visual inspection of the sensor and of the general area where the sensor is installed. The manufacturer's recommended testing procedures should be followed.

Within each area being monitored by sensors, the test should (1) ensure that the system meets the manufacturer's specifications for probability of detection, (2) verify that no dead spots exist in the zone of protection, and (3) verify that line supervision and tamper protection in both the access and secure modes are functional. Records of testing results and equipment sensitivity settings or voltage outputs should be maintained so that deterioration in equipment capability can be monitored. Walk tests should be performed for all areas covered by the sensor and compared with the results of the acceptance test to check for any degradation in the coverage of the sensor or misalignment problems. Significant changes in room configuration could affect sensor coverage. If room configuration has changed significantly, a complete performance test of the sensor coverage should be initiated to ensure asset or room protection. Because the PIR is the dominant technology in this configuration, the performance test guidelines described in the PIR performance testing section should be followed. These tests should answer the questions listed below:

- Does the sensor sensitivity decrease at higher room temperatures?

- Can the sensor be covered without generating an alarm?

- Can a person shield his body temperature from the sensor?

4.5.6.3   Operability Testing

The operability testing should consist of a simple walk test and tamper test on each sensor in the system. Step tests should also be conducted to verify proper sensor operability. The step test is performed starting at likely points of entry and along paths toward protected items or areas.

For example, in the positions numbered 1, 2, and 3 in Figure 56, operational tests are performed weekly. Locations 1 and 2 begin at likely points of entry—the door and window. Location 3 is an additional test near the door. The alarm stations are contacted before each test. The test subject remains still until the alarm station operator signals that the sensor is reset and is in the secure state. The test subject takes three steps into the room and stops. The alarm station operator verifies that an alarm from the sensor was received from the correct location. If any of these simple tests fails to initiate an alarm, the sensor should be checked for alignment problems.

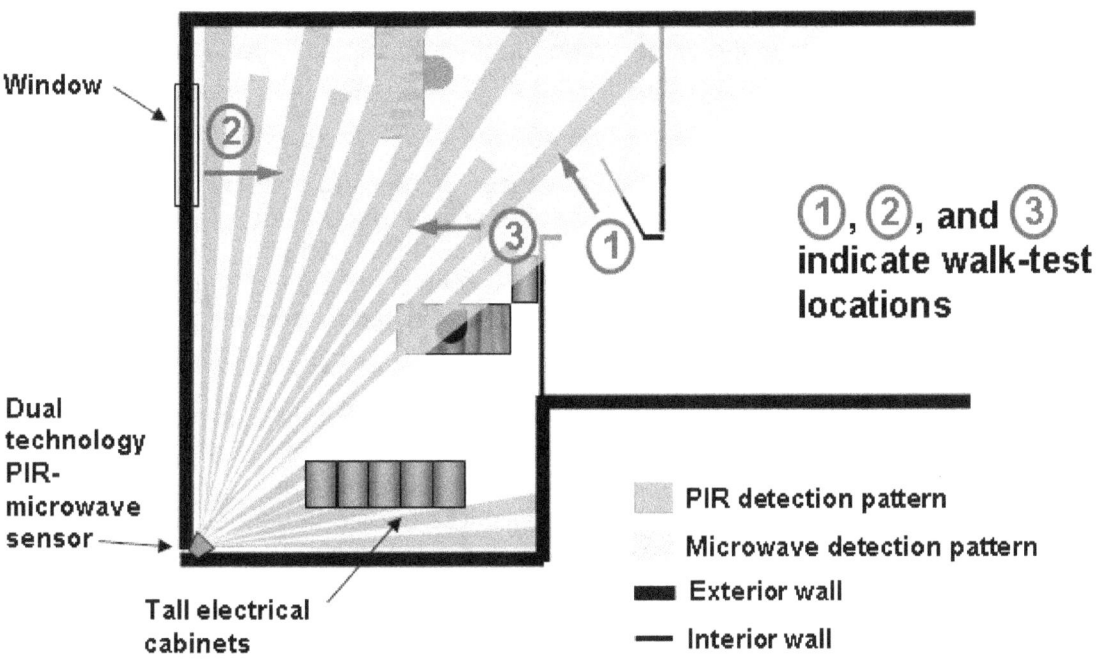

Figure 56: This diagram illustrates the possible walk test locations in a room with sensor coverage by a dual-technology PIR-microwave sensor. Note two things: (1) The microwave portion of the sensor can penetrate light construction, but since the sensor's "AND" logic requires that both sensors detect before an alarm is registered, a person in the adjoining room won't cause a nuisance alarm, and (2) the tall electrical cabinets block a portion of the sensor's detection pattern. This could be a problem if the electrical cabinets are a possible target and it is possible for an adversary to hide in the location where the detection pattern is blocked.

4.5.7 Maintenance

A visual inspection of the installation should be performed periodically, particularly after major maintenance to the building in the sensor area. Mounting brackets and hardware should be inspected for stability and corrosion. Frequent visual inspections ensure that no blocking objects have been moved into a position that would render the sensor inoperative. Periodic tests, in addition to self-test invoked by the sensor or the system, ensure that the sensor is operating effectively. Standby batteries should be replaced on a conservative schedule. Every service call should be entered in a log to record the date, time, corrective action, and an assessment of the cause of the problem.

4.6 Video Motion Detection

4.6.1 Principles of Operation

Video motion detection (VMD) can be added to either analog or digital camera systems. VMD has been effectively implemented using daylight cameras, near-infrared cameras, thermal imagers, and 360-degree-view cameras. VMD requires the addition of hardware modules and/or alarm processing hardware and software. The technology is modular so that it can be implemented at either the camera or at the alarm station. In one configuration, VMD software can be downloaded into specifically configured digital cameras with embedded digital signal processor (DSP) chips and memory so that the detection function occurs at the camera. When a VMD detection event occurs, the camera sends an alarm message to the alarm stations and then increases the frame rate of the video subsequently transmitted. Prealarm video can be stored in the digital camera's memory, and then the DSP transmits it to the alarm stations when an alarm event has occurred.

VMD technology has experienced significant advances in the state of the art since the early 2000 timeframe. Lower performance modules are available that provide simple movement detection, while higher performing equipment employs sophisticated algorithms to detect and categorize a moving target. With the exception of the "object left behind" algorithm discussed later, the VMD analysis algorithms are generally activated when movement occurs in the camera's field of view.

The technology makes decisions about what is moving and the nature of the movement occurring in the camera's field of view. Pixel movements are identified. Then the pixels in movement are "blobbed together" as a group of pixels in movement as a group. The blob of pixels is analyzed to determine if it falls into the classification criteria needed to generate an alarm. Then the motion, direction, and speed of motion (among other factors) are analyzed. If the attributes of the motion pass the algorithm analysis tests for being a valid intrusion motion, then an alarm signal is transmitted to alarm station's alarm display, and video of the intrusion event is displayed on an alarm station video monitor.

Early VMD equipment consisted of a module inserted in a camera's video transmission circuit. It highlighted the portion of the video image where motion was detected. Some units highlighted an area of the video image when a certain percentage of pixels in the camera's field of view changed. The video processing implemented in these early VMD modules was simplistic and, as a result, produced numerous nuisance alarms, thus making the technological enhancement not very usable for reliable intrusion detection purposes. Early VMD modules responded to essentially anything that moved, including insects walking on the camera enclosure's front cover glass. As the technology advanced, areas within the camera's field of view could be set as an active detection area ignoring movement in all other viewed areas. With increased sophistication of activity-related detection algorithms, different portions of the viewed area could be made active for different detection events. For example, one area could be set to alarm on any movement, while another area could be set to alarm when there was movement from right to left and not when the movement was from left to right.

In recent years, progressive VMD vendors have incorporated significant sophistication into the algorithms that analyze motion. Users can calibrate the camera's field of view with respect to object size for purposes of classification, object type determination (human, vehicle), speed, size, direction, and location. For example, during camera field-of-view calibration, one approach is to have a person stand in the camera's field of view at three or four locations from near field to far field of view. At each location, the operator calibrates the software to an individual's height. For each location, that height calibration information is saved for use by the detection and classification algorithms. The analysis software can then accommodate differences in object size with respect to location in the camera's field of view and distance from the camera. A human occupies many more pixels in the camera's field of view when closer to the camera than when further away. To compensate, VMD software can scale the detection algorithm's function for identifying human movement throughout the camera's field of view.

In some VMD equipment software, it is possible to calibrate the alarm algorithm to allow detection of movement at the ground level so that movement that occurs above ground can be exempt from classification as an alarm target. That assumes that detection of a person either walking upright or crawling on the floor is the target of interest. However, if there is a concern about a person swinging into an area on a rope, the robustness of a VMD system's detection algorithm would have to be tested to ensure that the equipment could detect the motion of a target in particular size ranges.

VMD detection and alarm notification are based on a set of "rules" and "areas of interest" defined by a system operator. This capability allows specific detection functions to be active only within certain portions of the camera's field of view. The detection function does not have to be implemented throughout the entire camera viewing area. For example, if the area of interest is inside a large room and there is an aisle for passage through the room alongside the area of interest, the intrusion detection function can be configured to be operational for the area of interest, while human movement outside the area of interest does not trigger an alarm.

VMD vendors have established many detection rules for their hardware and software configurations. The specific kinds of rules, their functionality, and their reliability vary widely from vendor to vendor and from application to application. The technology supplied by a vendor of interest should be thoroughly tested in the specific applications envisioned before committing to the installation of a particular technology.

4.6.2 Types of Video Motion Technologies Available

VMD functionality is available in three configurations:

(1)   Software running on a personal computer with video capture cards

(2)   Stand-alone single- or multi-channel hardware/software modules

(3)   Software embedded within a digital camera with an onboard DSP chip and associated memory

The first configuration is normally at the video head-end located at or near an alarm station. Either analog or digital cameras can be connected to the VMD computer, depending on a particular vendor's camera options. The second configuration can be located at either the camera or alarm station location. Either analog or digital cameras can be connected to these VMD modules. The third configuration is located within the digital cameras at the camera locations.

Some vendors have advanced VMD capabilities to include a tracking function. A movable (pan-tilt-zoom) camera is set to view a static scene. When alarm-generating movement is detected, the camera moves and zooms in on the target and tracks the target within the limits of the camera's movement and zoom capabilities. The function has been applied to static or preset locations in a camera tour of several fields of view, and the tracking function can be triggered at any one of the camera's tour stop locations. At least one vendor has VMD equipment with the ability to track movement while the camera is in motion. While the camera is panning, the VMD software identifies relative movement within the camera's field of view and then proceeds to track that movement.

Electrical connection of cameras to VMD equipment is simple and straightforward. Analog cameras are connected to a VMD processor using coaxial cable Bayonet Neill-Concelman (BNC) connectors, while digital cameras are connected using Ethernet connections to a network of high-bandwidth digital switches. High-bandwidth digital networking and trunking to support digital video transmission between camera locations and the alarm stations are beyond the scope of this discussion. However, it is instructive to note that transmission of high-bandwidth video signals is, in many cases, not viable using Ethernet systems designed for message transmission such as e-mail and Internet access. Specialized network configuration expertise is required for the design and installation of a digital Ethernet network to support efficient and reliable high-bandwidth video transmission.

4.6.3 Sources of Nuisance Alarms in Video Motion Detector Systems

Commercial VMD technologies have varying degrees of performance; however, performance testing of the technology has identified conditions that produce performance challenges. Generally speaking, indoor environments tend to produce significantly fewer challenges than do outdoor environments. Large changes in lighting conditions, reflections from shiny objects, shaking cameras, out-of-focus cameras, a low-contrast scene, a target color near the same color as background, a target not occupying enough pixels in the field of view, movement of large items in the field of view, such as trees or large birds, and heavy, blowing snow and driving rain have been identified as sources of nuisance alarms. Available equipment has a wide range of intruder detection and nuisance alarm performance attributes. Therefore, it is necessary to thoroughly test the VMD equipment to ensure that the equipment performs to expectation before system purchase and installation.

When used indoors where environmental variables are significantly fewer, VMD technology produces significantly fewer nuisance alarms. Indoor lighting in most locations is fairly constant throughout the day and generally the camera-to-target distance is much shorter than that encountered in outdoor applications. Cameras tend not to shake and vibrate in indoor applications, and animals are not present to trigger alarms.

4.6.4 Characteristics and Applications

Combining the use of VMD with assessment cameras in indoor applications provides sensor functionality without the use of a physical sensor. VMD applications do not have the phenomenology associated with a physical sensor to create a detection alarm. The camera and VMD software are not sensing the presence of an intruder within a sensored space. Changing attributes of a video image are being analyzed by software, and the results of that analysis determine if an alarm condition is present. Current video analysis software only approximates a portion of the detection and assessment capability of the human mind. While VMD detection software has significantly improved since 2000, VMD software is definitely not superior to human visual acuity and cognition. However, VMD software does provide surveillance 24 hours a day, 7 days a week, to respond to predefined targets and attributes of movement within a scene. Humans do not have the capability to continuously focus on a scene for extended lengths of time. VMD provides that continuous observation and alerts the alarm station operator to allow a human to make the final decision regarding the presence of an intruder.

As with a physical sensor, VMD provides an indication that there is a change in the area under observation when an alarm is generated. The change creating the alarm is visual rather than phenomenological. Diagnostic information is also provided to the operator with a VMD alarm that is not provided with physical sensor alarms. The VMD software puts a box around the identified target in the assessment video to draw the operator's attention to a particular location in the video scene. The operator can then focus on the box in the video to assess what triggered the VMD alarm.

When designing for VMD sensor installation, knowledge of preferential sensitivities associated with technology should be understood. The technology is more sensitive to movements across the camera's field of view. Movements toward or away from the camera are less sensitive to detection. Movements across the camera's field of view change more pixels in the image with the same amount of movement than does movement towards or away from the camera. Therefore, when VMD is used as a detection sensor, the camera should view the scene so that the detection pattern is across the camera's field of view. VMD should not be used with cameras that view scenes that experience large changes in scene illumination. For example, if a camera is oriented so that it views the inside of an exterior door, opening the door on a bright sunny day will cause a large change in scene illumination when the door is opened and may either cause a nuisance alarm or nondetection of personnel entry through the door because of the significant perturbation presented to the classification and detection algorithms.

VMD tends to be more sensitive to black pixel and white pixel movements. These colors are at the extremes of the color spectrum and less decisionmaking is needed about pixel movement as compared to movement of pixels in the middle of the grey scale or with muted colors. Because of the sensitivity to white and black pixels, the scene's area of interest for VMD detection must not include an area that experiences moving shadows or moving sun glint. If VMD is applied in an area where there are windows (particularly if the windows are facing east or west), the low sun angle entering the windows can cause persons to cast shadows inside the building. If the area of interest includes an area where moving shadows are cast, this may cause unnecessary activation of the detection and classification algorithms creating the possibility of nuisance alarms.

The use of VMD with cameras that shake or vibrate is problematic. From the VMD software's perspective, all the pixels in the camera's field of view appear to be in motion. This results in significant image processing and analysis. If there is by chance some patterned motion within the random motion of all the pixels in motion, the detection algorithm will create an alarm because, in its analysis, there appeared to be "motion with purpose" in the video stream.

As described earlier, some VMD software algorithms can adapt to slow scene changes. This adaptive response is needed, particularly for exterior applications where outdoor illumination in lighted areas may vary by four to five orders of magnitude. Adaptive algorithms can compensate for very slow movements. A very determined intruder could gain entry into an area with very slow movement. Also, some VMD systems are not sensitive to movements of pixels of a color similar to the background color. If this is the case, intruders could cloak themselves in fabric of a color similar to the background or floor and very slowly gain access.

Similarly, if a thermal imager is the video source for VMD analysis, an intruder could insulate themselves with highly insulating garments and then cloak themselves under another thick insulating material so that the thermal imager does not view movement of a human because everything in the thermal imager's field of view appears to be at the same temperature. In these cases, complementary sensors with orthogonal detection criteria would be needed.

4.6.5 Installation Criteria

As described in the nuisance alarm discussion above, cameras for VMD should view a detection area so that intruder movement is across the camera's field of view rather than up or down. Cameras should be installed so that they are looking at an area of interest at an angle. Mounting cameras on a ceiling so that they are looking straight down is problematic for some VMD algorithms because looking straight down at the floor causes the top and bottom of the camera's far field of view to be at the same distance from the camera. Algorithms are normally expecting that the top of the camera's field of view be further away from the camera than the bottom.

Scene background should be a neutral color rather than a very light or very dark color. A scene with a very light or very dark background color makes it easier for an intruder to blend in with the background.

Cameras for VMD sensors should be installed on solid structures that do not shake or vibrate. This minimizes the amount of computations that the VMD algorithms must make and improves the ability to detect an intruder.

For indoor locations, installing visible light cameras with area illumination rather than thermal imager cameras is preferable from a detection perspective.

VMD detection areas should not be implemented in locations where shadows move across the detection area. Because of VMD detection algorithm sensitivies, unnecessary alarms can be generated by moving shadows in a camera's field of view.

VMD cameras should not be installed so that they view the inside of exterior doors. Opening a door during bright sunlit days will cause large bright spots in the camera's field of view, which may cause the VMD algorithm to not detect human movement in that area.

VMD technology should be installed so that it is integrated with either a digital video recorder or network video recorder for instant playback of the alarm video. When the camera video images are processed through a VMD module or processor before being recorded on digital media, the recorded video will be "marked up"; that is, the recorded video will have alarm boxes around the area where the VMD camera observed movement. This allows the operator to focus on the portion of the video image where movement was detected rather than having to search for that location in video that is not marked up. Most VMD cameras can transmit and produce images to be monitored at monitoring locations such as alarm stations within 2 seconds from detection which is generally an acceptable standard of performance.

4.6.6 Testing

Testing a VMD sensor is very similar to testing a physical sensor. Targets of interest move through the scene at a range of speeds and at varying aspect ratios. If a human is the target of interest, movement includes the following:

- Walking and running at various speeds
- Walking and crawling at normal and extremely slow speeds through a VMD camera's field of view to ensure that an alarm is created

Tests should also include a human doing a belly crawl with a fabric cover that is approximately the same color as the background or floor.

A regular program of testing sensors is imperative for maintaining them in optimal operating order. Three types of testing need to be performed at different times in the life of a sensor: acceptance testing, performance testing, and operability testing.

Alarms should be produced deliberately during all types of tests.

4.6.6.1   Acceptance Testing

Acceptance testing for the video assessment system is the most encompassing because baseline performance and operability are determined and documented. Acceptance tests will uncover operational and functionality issues that need to be addressed to ensure system operation in accordance with design specifications. Acceptance tests should be performed to ensure that the VMD software produces alarm messages in response to tamper or defeat attempts. Use the following guidelines for acceptance tests:

4.6.6.1.1   Testing the Camera-to-Alarm-Station Connection:

(1)   Ensure that each camera produces a VMD video image on the monitor.

(2)   If camera images have graphic legends displayed on the monitor, ensure that the graphic legends are labeled correctly for the VMD camera channel being tested.

(3)   Ensure that VMD intrusion alarms in assessed areas trigger the appropriate camera's video to appear on the alarm station monitor. For example, an individual in the field communicating with an individual in the alarm station using a two-way radio triggers an intruder alarm in each VMD-sensored zone and ensures that VMD video resulting from the alarm has a box around the intruder and is from the correct camera for the zone in alarm. This individual also ensures that the assessment video appears on the alarm station monitor within the timeframe as specified by the manufacturer (2 seconds is generally an acceptable standard) and the playback of recorded video is from the correct camera.

4.6.6.1.2   Testing for the Correct Error Message by Simulating Potential Problems:

(1)   Disconnect power to individual cameras at field camera junction boxes, and ensure that a VMD "loss of video" alarm occurs for the camera disconnected.

(2)   Cover the front of camera enclosure with a black plastic bag or densely woven cloth, and ensure that a VMD "loss of video contrast" alarm occurs for the camera tested.

(3)   Shine a bright light into the front of each camera, and ensure that a VMD "loss of video contrast" alarm occurs for the camera tested.

(4)   Turn off the lights in the area viewed by the camera to produce a dark, low-contrast image, and ensure that a VMD "loss of video contrast" alarm occurs for the camera tested.

(5)   Disconnect the video signal cable from each analog camera, and ensure that a VMD "loss of video" alarm occurs for each camera channel disconnected.

(6)   Disconnect the fiber-optic video transmission cable (fiber) for each camera, and ensure that a VMD "loss of video" alarm occurs for the camera channel disconnected.

(7)   Physically move or rotate the camera so it is no longer viewing the intended scene, and ensure that it generates an error message.

4.6.6.1.3   Testing for Proper Ethernet Connections:

(1)   Disconnect the Ethernet cable from each digital camera, and ensure that a VMD "loss of video" alarm occurs for the camera channel disconnected.

(2)   Disconnect the camera Ethernet cable to Ethernet switches carrying video signals, and ensure that a VMD "loss of video" alarm occurs for all cameras connected to the switch.

(3)   Disconnect the power cable to Ethernet switches connecting video signals to the VMD hardware, and ensure that a VMD "loss of video" alarm occurs on all camera channels connected to the switch.

4.6.6.1.4   Testing for Multiple Intrusion Alarms in Multiple Zones:

(1)   Simultaneously create intrusion alarms in two, three, four, and five adjacent and nonadjacent perimeter zones, and ensure that the alarm assessment video queues up the recorded (prealarm) video and that the recorded video is displayed for each of the alarming zones.

4.6.6.2   Miscellaneous Testing

In areas where sunlight enters the area, during daylight periods, an observer should ascertain that alarms are not created as the sunlight and the shadows it creates moves across the VMD-sensored area. Also under the same conditions, an observer should ensure that shadows created by persons walking adjacent to VMD-sensored areas do not create alarms.

At six-month intervals, secondary power source operation should be tested by disconnecting the primary power source to the security system and ensuring that the security system continues to function without creating detection alarms or system failure.

4.6.6.3   Performance Testing

VMD performance is also affected by the issues indicated in Section 4.6.7.4, "Operability Testing." However, performance degradation is usually caused by changes of lower severity in the issues noted above. Usually, the performance or functionality degrades but the functionality does not cease completely. Physical changes in the baseline scene will also affect VMD performance. For example, if a piece of furniture or equipment is permanently added or removed from the scene, the baseline scene must be recalibrated to reset the detection algorithms to the new baseline condition. Some VMD software readily adapts to changes in the baseline scene over some period of time. However, to ensure proper performance, a manual recalibration of the camera viewing the changed scene should be performed.

To optimize digital network bandwidth utilization, digital video processing at the camera generally involves video image compression. Video compression reduces scene resolution to minimize the amount of information transmitted from the camera to the alarm station. The amount of compression applied affects the amount of fine detail taken out of the video images. Application of greater than 20-percent compression begins to significantly affect fine detail in the transmitted image. Performance degradation with increased compression is noticed initially at the camera's far field of view and, as compression is increased, that degradation begins to affect more and more of the camera's scene toward the camera. Compression causes the edges of objects to become fuzzy. Because VMD software analyzes the edges of groups of pixels in movement, compression that causes images to become fuzzy significantly complicates and degrades the functionality of the algorithm analysis.

4.6.6.4 Operability Testing

VMD operability issues are generally a result of camera degradation or failure, video transmission system degradation or failure, video system power or secondary power source failure, system grounding or ground loops, or the result of lightning-induced events.

Operability issues can also result from changes to the detection rules for a VMD scene or changes to the VMD calendar or time clock scheduler.

Physical changes affecting VMD operability include out-of-focus camera; camera moved from its initial VMD calibrated position; loss of scene contrast; degradation or failure of camera electronics; loss of, or significant change in, area illumination; or bright light appearing in the camera's field of view.

4.6.7 Maintenance

Specific maintenance for VMD should focus on maintenance of the camera and video transmission system. Because the heart of a VMD sensor is essentially software that analyzes camera video, it is anticipated that the software will be undergoing continuous upgrading as the state of the art for this technology advances. Therefore, it is anticipated that the VMD software may require updates during the life of the VMD sensor.

The camera lens and enclosure front cover glass should be cleaned and the lens focus adjusted, as necessary.

On a yearly basis, each camera's image output should be viewed at both alarm stations to ensure that the picture is clear and crisp and is not washed out or of low contrast. The absence of a clear, crisp picture would indicate degradation in the camera's electronic circuitry, or in the case of analog cameras, it could indicate degradation in the video transmission network or video amplifier modules.

In addition, the tests shown above in the section on acceptance testing should be performed to ensure that all the alarm security features are still active and operate properly.

Uninterruptible power supply batteries for modules providing backup power to the security system should be replaced at 3-year intervals.

5. WATERBORNE SENSORS

5.1 Waterborne Sensor Systems

This section provides information about waterborne sensors for facilities with water intakes, as illustrated in Figure 57.

Figure 57: An example of a fiber-optic grid covering a water intake channel.

5.1.1 Principles of Operation

A common application for intrusion detection in a waterbourne environment is the use of fiber-optic grid sensors. Fiber-optic grid sensors can be made to accommodate various waterbourne applications and configurations with minimal impact to the operational capabilities of the structures or systems to which it is attached. When evaluating the detection capabilities of fiber-optic grid sensors, the security industry measures quality by sensing bends and breaks in fiber-optic grids. Typical security configurations use single-mode diameter fiber illuminated with approximately 1,500-nanometer (nm) wavelength light, but any fiber diameter or signal wavelength are acceptable as long as bends and breaks can be sensed quickly (typically within a few seconds or less, but the minimum requirement depends on the overall security system configuration). Maximum fiber length typically is a few kilometers, but some vendors claim maximum lengths of many times that. What constitutes a "bend," resulting in signal attenuation without total signal loss, will depend on characterization of potential nuisance alarms (e.g., if the fiber commonly bends in a strong wind, that will result in more typical attenuation than a fiber run indoors, and the sensing system should be adjusted accordingly).

Once the fiber grid is arranged to ensure that an adversary must break it to enter the facility being protected, the fiber is connected to sensing electronics. These electronics check fiber continuity, providing an optical signal into the fiber and measuring its output on the other end and ensuring that the signal is not greatly attenuated or removed altogether. Some sensing systems are capable of detecting where in the fiber line the event is occurring, in addition to the event itself, but this capability costs more and is often not necessary for adequate protection.

Most fiber sensing systems will close or open a dry contact closure relay upon sensing a bend or break event, but output is adjustable depending on application needs.

Fiber-optic cable is especially preferred by security professionals from a tampering perspective as its modification requires very specialized tools and delicate procedures. Electrical wire, on the other hand, typically just needs a few clips and an extra loop of wire to create enough slack to penetrate without detection.

5.1.2 Types of Fiber Grid Available

In protecting an area or surface, several vendors have had success creating fiber nets, and some have embedded the fiber in a plastic or metal grate for added strength. The increased rigidity adds confidence that any break in the fiber is intentional, not accidental or caused by the environment.

- A ***fiber net*** can be easily wrapped around an object or surface being protected, or stretched taut across an opening that needs to be controlled. Because of its flexibility, fiber net is easier to disturb and bend, requiring very little force to cause an intrusion event.

- A ***fiber grate*** is a rigid structure typically installed across openings to be controlled, and any break or bend sensed is almost always caused by destruction of the surrounding medium; the fiber detects that the rigid structure is being breached.

Both types can be treated to prevent biofouling (underwater growth including algae, barnacles, or simple electrochemical corrosion) if deployed in a marine environment. (See Figure 58.)

Figure 58: This fiber-optic grid was tested under water for about 6 weeks and shows biofouling through barnacles, which can eventually degrade the integrity of the materials.

5.1.3 Sources of Nuisance Alarms

Of the two major types of fiber grid, the more flexible fiber net experiences multiple nuisance alarms in a marine environment. Marine mammals are often inquisitive and can bump the net, which results in enough signal attenuation to cause an alarm. A strong water current flowing through the net can distort it, also causing an alarm. The sensing system threshold can be adjusted to allow these types of events, but at the cost of possibly not sensing actual malicious bending and stretching of the net. Also, exact characterization of marine mammal interference is difficult.

The fiber grate, a much more rigid structure, is typically considered free of nuisance alarms.

5.1.4 Characteristics and Applications

The fiber net can be used on a shaped surface that needs to be protected, an odd-shape opening that needs to be protected quickly, or a location where weight is a major consideration. A fiber net can also detect activities such as climbing and is, in general, more responsive (than a grate) to environmental influences. Fiber nets work well for detection of several types of events (i.e., pushing, climbing, gathering, cutting, and others). A net is relatively cheap and easily replaced.

A metal or plastic fiber grate creates delay (as long as an alarm is generated) and strong access denial and works where the primary activity to be sensed is cutting. Because a grate is a rigid structure, a single cut does not grant access but will sound an alarm, which allows notice of malicious activity before adversaries are able to pass through the barrier. A "smart fence" can tell when someone has broken through it, and it can be manufactured in any reasonable size.

The following characteristics apply to fiber nets:

- Fiber nets are weaker than grates and are more vulnerable to damage in turbulent environments or where heavy debris could be carried into them.

- Fiber nets can be slashed, so swimmers are able to pass through the net immediately.

- Fiber nets can be "noisy" (have frequent nuisance alarms) caused by underwater currents flowing against them or from marine mammals bumping up against them.

The following characteristics apply to fiber grates:

- Fiber grates are inflexible, so if a swimmer were to cut the grate free from its support structure, it could be completely lifted away without an alarm being sensed (adding antitamper devices can mitigate this problem).

- Fiber grates can corrode quickly if metal grates are not installed correctly (with a zinc or magnesium sacrificial bar, which must also be maintained and/or replaced during the life of the grate).

- Plastic fiber grates can embrittle. Low temperatures and exposure to ultraviolet light can embrittle plastic grates.

The following characteristics apply to both fiber nets and grates:

- Fiber grids are susceptible to biofouling, resulting in corrosion and damage. Organisms such as barnacles could restrict water flow through a grate.

- Regular cleaning may be necessary to prevent interference from barnacles. (See Figure 58.) Anti-biofouling paints are effective against organisms adhering to the grate but are usually toxic to the environment.

- Fiber grids require assessment to confirm an alarm.

- Fiber grids are difficult to service underwater.

5.1.5 Installation Criteria

When installing either fiber grid system, it is necessary to work closely with the manufacturer to ensure that all potential problems are mitigated. Installation criteria include the following:

- Install a sacrificial zinc or magnesium block on metal components.

- Test plastic material in the ultraviolet lighting conditions or temperature ranges experienced in the location to be installed.

- Use anti-biofouling coatings (if the wildlife in the area will not be affected) and determine cleaning methods and schedules.

- Add tamper-indicating conditions to ensure that the grate or net cannot be defeated without causing a tamper alarm or from being circumvented completely. For example, ensure that it is difficult to make a new opening next to a secure opening.

5.1.6 Testing

A regular program of testing sensors is imperative for maintaining them in optimal operating order. Three types of testing need to be performed at different times in the life of a sensor: acceptance testing, performance testing, and operability testing.

5.1.6.1 Acceptance Testing

Acceptance testing should answer the following questions:

(1) What is the vendor's prediction for long-term survivability of the grid? While some corrosion or decay is expected over time, will the grate or net degrade in a timeframe that is acceptable to the facility?

(2) What strategies will be employed to prevent biofouling? Will the grate or net become overgrown with algae or barnacles?

(3) Could the grate or net be accidentally damaged, and what strategies will be used to prevent accidental damage?

(4)   What is the projected maintenance or replacement schedule and what are the associated costs?

(5)   What are the tamper-indicating strategies employed for the grid? Does the system alarm when fiber is unplugged? Testers should simulate a fiber disconnection instead of actually unplugging the fiber because the fiber connector ends must be cleaned before re-connecting.

(6)   What is the acceptable threshold to prevent frequent nuisance alarms? Does the system alarm if the grate or net is rattled or hit?

(7)   Does the grid respond to a break? Work with the manufacturer to simulate breaking the grid (either grate or net) to generate an alarm instead of actually breaking the materials.

5.1.6.2   Performance Testing

Performance testing should be done regularly to ensure that the system meets performance criteria, as follows:

(1)   To ensure that the system will not report frequent nuisance alarms, shake or hammer the grid to simulate a marine mammal or water current interference and ensure that no alarm is sensed.

(2)   Based on the manufacturer's recommendation, simulate breaking the grid (either grate or net) to generate an intrusion alarm instead of actually breaking the materials.

5.1.6.3   Operability Testing

Operability testing ensures that the system works as intended. Operability testing for fiber-optic grids involves nondestructive testing. Since the primary event to be sensed is cutting, the primary test will follow the vendor's prescribed methods to simulate cutting.

5.1.7 Maintenance

Maintenance should be performed according to a defined schedule. If performance or operability testing shows degraded system performance, the maintenance schedule should be adjusted accordingly. The following maintenance is recommended (if not already required by the vendor):

• Clean the grid, or verify that there is no growth on the grid.

• Clear any debris from the grid mesh.

• Check for corrosion around signal lines, connectors, and the sensor structure.

6. VIDEO ASSESSMENT

6.1 Video Assessment Systems

6.1.1 Overview

Assessment is a critical component of detection and is equally important to the initiation of response. Assessment provides a means to determine the necessary actions (responses) needed to mitigate situations that pose a challenge to physical security. The key element within the assessment process is identification. Identification assists the security force in selecting appropriate responses within a force continuum to address security-related and potential threat situations resulting from the detected activity. Likewise, identification provides a means for the security force to determine the absence of a threat resulting from detected activity such as a nuisance alarm caused by wildlife or debris. It is equally important that the assessment techniques identify the stimulus that caused the alarm quickly, before the stimulus of the alarm disappears from view. This enables the initiation of timely response consistent with the goals and objectives of the physical protection program and protective strategy. Therefore, video assessment systems should be robust and capable of providing the highest level of protection for the specific application in which they are employed.

6.1.2 Principles of Operation for Video Alarm Assessment Systems

A video alarm assessment system allows security personnel to determine rapidly, after a sensor alarm, if an intrusion has taken place within the facility. In addition to real-time observation of the location where the sensor alarm occurred, the video assessment system can record and store video of alarm events and other significant event information. This permits retrieval of event images within the protected area, even under conditions of multiple simultaneous alarms or delayed security personnel attention.

The design of the video assessment subsystem should consider the following goals:

- Ability to completely assess all sensor locations

- Ability to assess nuisance alarm sources

- Ability to provide system response fast enough to assess the area before the intruder leaves the area

A video alarm assessment system consists of cameras at sensored areas that display video images on monitors in the alarm stations, video switching, and recording and communication systems for transmission of the video images from the cameras to the alarm station monitors. Figure 59 and Figure 60 show block diagrams of analog and digital alarm assessment system components. The major components of the video assessment system include the following:

- Cameras and lenses to convert an optical image of an assessed scene into an electrical signal

- A lighting system to illuminate isolation zones evenly and sufficient intensity to produce usable video images of assessed areas at night

- A video transmission system that connects the cameras to the alarm station switchers and monitors

- A video switching system that takes the video from cameras and routes it to recording and monitoring display devices

- A video recording system that records video from camera feeds for archival needs or alarm event recording

- Video monitors for display of alarm assessment video

- An alarm communication and display controller to interface between the alarm sensor system and the video alarm assessment system to display video in the alarm stations when a detection sensor detects activity and produces an alarm

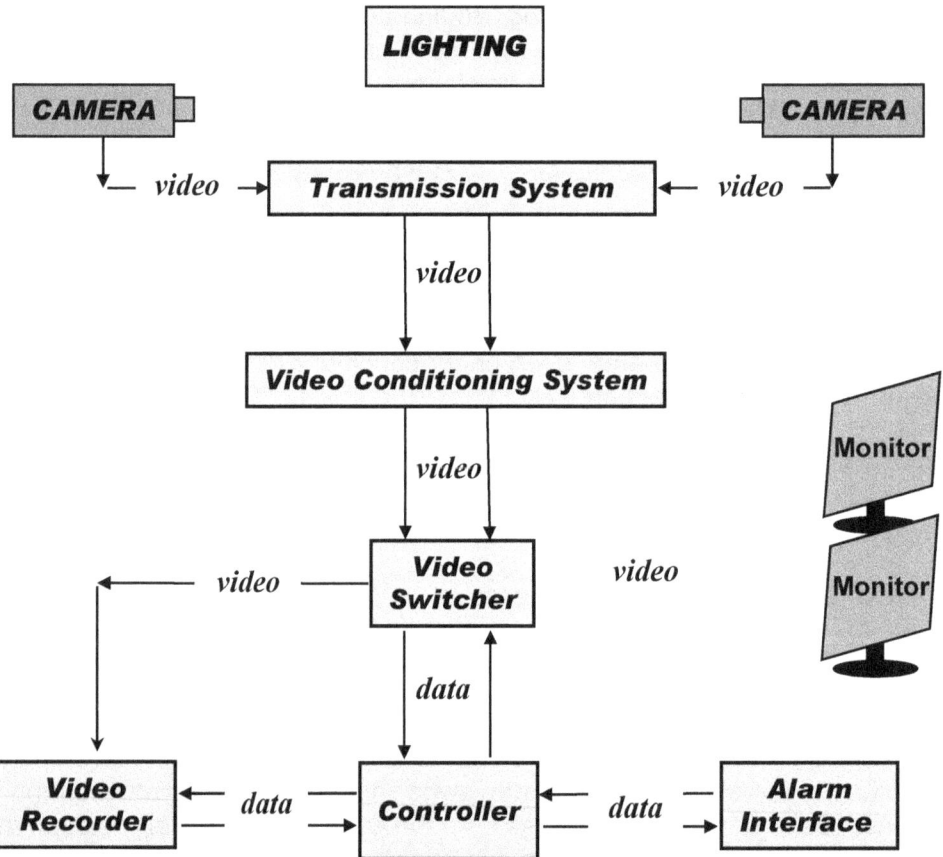

Figure 59: Diagram of an analog video system.

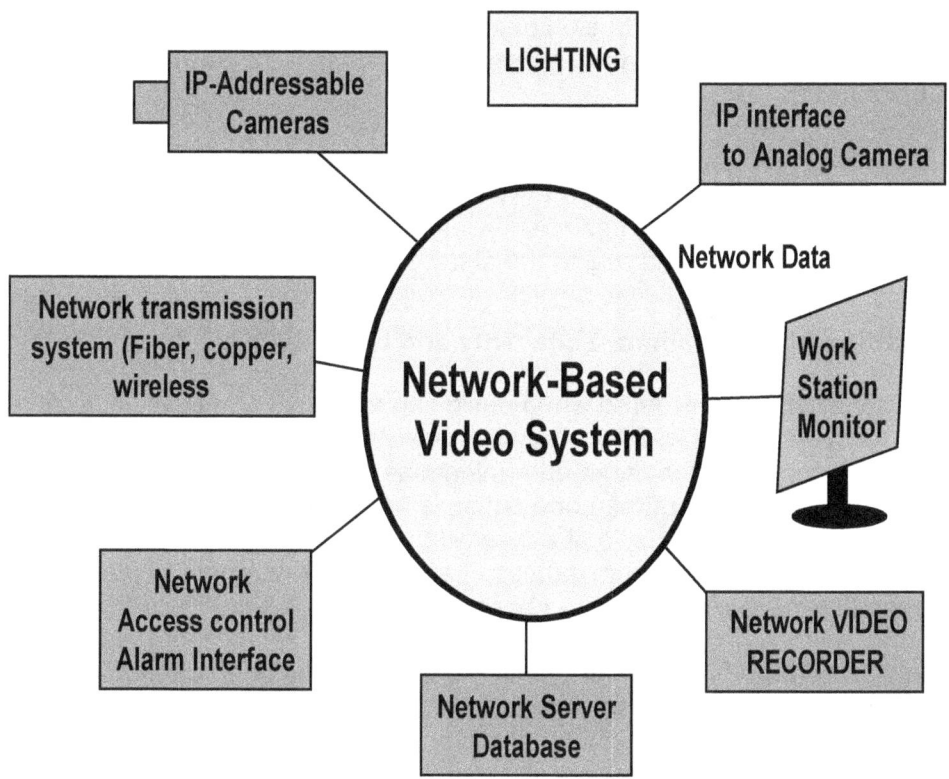

Figure 60: Diagram of a network-based digital video system.

A camera converts the physical scene in the camera's field-of-view into an electrical signal suitable for transmission over a cable, fiber, or wireless media to a video monitor for viewing by an observer and digital video recording for playback and archival information. The video switcher directs the signal to a selected monitor for display of the visual scene observed by the camera. A selected signal can also be routed by the video switcher to and from a video recorder for playback of pre- and post-alarm video and for permanent archiving and forensic analysis.

Camera resolution, sensitivity, lens field of view, and aperture should be considered during the design of exterior perimeter isolation zone lengths and widths. Each video assessment system aspect should be considered individually and within a system. If assessments need to be made from recorded data, then the resolution of the recording device should be considered. To compensate for reduced video recorder playback resolution, reduced spacing between cameras and reduced maximum assessed area lengths may be required. The video recording and storage device may be the resolution-limiting factor in the assessment system. If video information is to be retrieved from archive recording, then the archive recorder resolution should be considered in the system design.

Additional assessment cameras, associated circuitry, and hardware must be installed if sensor zones need to be shortened by terrain variations or locations of personnel access points. Each sensor zone should be evaluated individually to ensure that appropriate assessment equipment is installed and assigned to provide optimal performance of assessment capabilities within the isolation zone. It is more beneficial to install multiple assessment cameras in an extremely long sensor zone than risk a reduced or limited assessment capability and thus reduce the overall level of protection provided by the intrusion detection and assessment system. Sensor zone width also has an effect on the maximum assessment zone length; narrower widths allow assessment of longer zones.

Most of the head-end video assessment equipment can be located in or near an alarm station for convenience, a temperature-controlled environment, and ease of access for maintenance. Normally, only the cameras, towers, power supplies, video transmission equipment, and signal transmission media are located external to the alarm stations.

6.1.3 How Lighting Affects Camera Sensitivity and Resolution

Camera sensitivity must be considered when designing the lighting system. A high-resolution camera will not function properly if the illumination is inadequate. The lens relative aperture (i.e., f-stop) must be considered in these determinations. Generally, faster lens speeds (lower f-number) are selected when more than one option is available, even if a lens with a lower f-number is somewhat more expensive. Lenses with f-numbers higher than f/1.8 should not be used except under very unusual circumstances. Lenses with f-numbers higher than f/1.8 will require increased lighting levels or a more sensitive camera.

A camera's sensitivity or performance under low light conditions often cannot be judged from manufacturers' specifications alone because a common standard does not exist. Some manufacturers reference scene illumination levels or illumination levels at the camera's imager that do not represent camera performance in security applications where a lens and its aperture affect the amount of light appearing on the camera's imager. These variant light levels are subjective and therefore do not allow performance comparisons. A side-by-side evaluation of cameras operating in the intended environment is the preferred way to determine a camera's suitability for assessing a particular area.

For a scene to be visible to the camera, it must be illuminated by sufficient natural or artificial light and must reflect a portion of that illumination for observation by the camera through a lens with appropriate field of view for assessment purposes. The light reaching the lens must then be clearly focused onto the camera's light-sensitive imager surface and converted into an electrical signal by the camera electronics system for transmission to the alarm stations.

6.1.4 Concepts of Cameras Used in Video Assessment

The most efficient use of assessment cameras is achieved when smaller camera imager size and longer focal length lenses are used. To the extent that these factors can be optimized, the number of cameras, towers, and camera subsystem infrastructure video transmission system components can be minimized. Longer focal length lenses have narrower fields of view. This minimizes the amount of area outside the perimeter being viewed. Assessment cameras can view more than one sensor zone in some circumstances. If a video assessment system is judiciously designed, when a camera outage occurs, an adjacent zone's camera may have the capability to assess a neighboring zone until that zone's camera is operational again.

Alarm assessment using video technology employs cameras and lenses installed in camera enclosures and accessory equipment to create video images for an operator to observe in a facility's central alarm station (CAS) or secondary alarm station (SAS). Cameras are manufactured to widely varying requirements, performance standards, quality levels, and reliability.

Analog security cameras are an offshoot of broadcast television technology; they are designed to provide a video output signal that is almost identical to that of an analog television camera. Currently, many analog camera vendors offer a wide range of camera technologies at widely varying prices. As camera technology advanced, the digital video camera became available. Many digital cameras are available to implement assessment systems using digital camera technology and digital video transmission systems.

The camera's electrical video signal is carried by a transmission media coaxial cable or fiber-optic cable to the alarm stations where it is displayed on a video monitor for viewing by an operator. Analog cameras produce an analog output signal in a specified format for viewing on a picture-tube-type or flat screen digital monitor. Digital cameras are connected to a computer network and provide a signal to the alarm stations through the network via a stream of digital packets similar to the process that produces video on a desktop computer.

6.1.5 Testing Video Assessment Systems

The testing portions of this section contain testing information that should be approached from a systems point of view. The testing section addresses several kinds of testing, including acceptance, operational and performance testing, integrating cameras, digital video recorders (DVRs), lighting, and system elements.

6.2 <u>Cameras</u>

6.2.1 Principles of Operation

Depending on the specific camera imager technology implemented, camera imagers respond to visible and near-infrared illumination or thermal signatures of targets within the camera's field of view. Camera placement depends on appropriate lens selection to adequately assess a defined area, zone, or sector. Detection sensor area and camera field of view area must be coincident. If a camera's field of view does not completely encompass a sensor's detection area, additional cameras must be deployed to ensure that all of the detection area can be observed with the video assessment system. Similar to motion picture format, a video picture comprises 30 still pictures per second taken in succession, compared to 24 still pictures per second in motion pictures. Each picture is called a frame.

Cameras have six distinct categories: black and white, color, day/night, infrared and infrared enhanced black and white, intensified, and thermal. Cameras with different technologies can be used together to provide a wide assessment capability for specific applications. This is particularly true when assessment must be performed in extremely low light conditions of less than 0.01 f-c (0.1 lux). For example, in unlit areas, a color camera can be used during the day and a thermal imager camera used at night.

The heart of a camera is a solid-state imager. The imager has a pixel array that converts light energy to an electronic charge. The performance of imager technology is affected by factors such as resolution, sensitivity, color versus black and white, and infrared or thermal spectrum sensitivity. The resolution of a quality analog color camera is approximately 470 lines. The resolution of a quality analog black and white camera is approximately 570 lines. Digital cameras have 520 to 4,000 lines of resolution. Infrared-enhanced camera imagers can provide video images appropriate for video assessment with low-power infrared illuminators. Thermal imagers have a range of 160 to 480 lines of resolution.

6.2.1.1 Camera Resolution

Resolution is the degree to which one can see fine detail in a viewed image. It is also a measure of spatial frequency or the number of pairs of alternate black and white evenly sized lines that can be seen in a given linear distance, typically expressed in line pairs per millimeter. The line pairs designation is used primarily in the field of optics, but the term also appears in television literature. Different camera resolution charts have been developed for color and for black and white cameras. Figure 61 shows examples of both color and black and white resolution charts.

Camera resolution is measured using a resolution chart where groups of equally spaced black and white lines arranged in a wedge-shaped pattern form the basis for resolution measurement. As shown in the left picture of Figure 61, the chart is normally calibrated for horizontal resolution between 200 and 1,600 horizontal TV lines (HTVL). The units of measure are stated as either horizontal or vertical TV lines. Vertical resolution is fixed by the scan rate. A color camera resolution chart is shown at right.

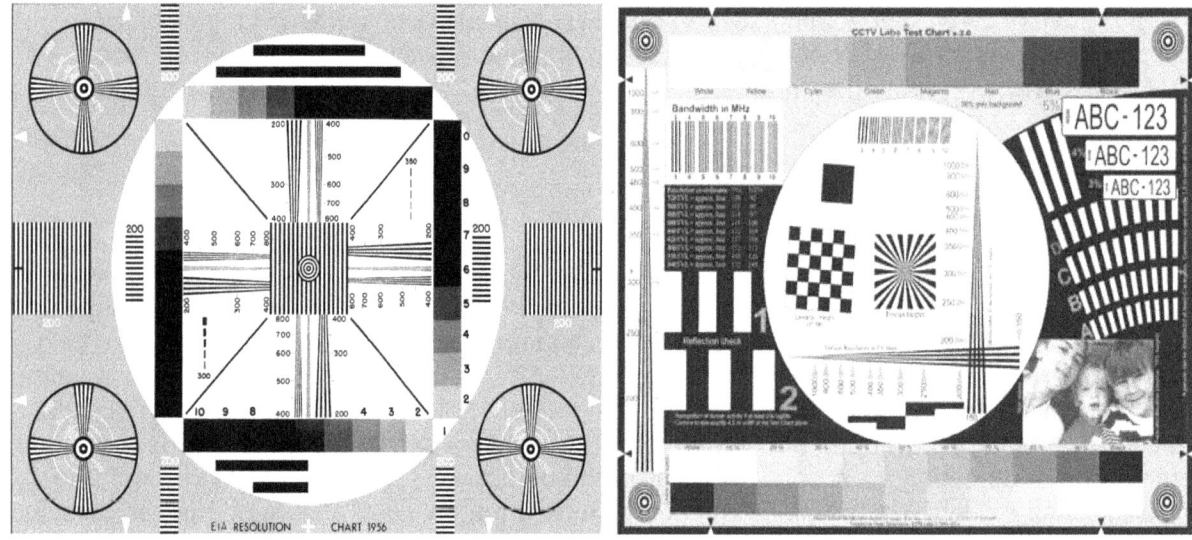

Graphic courtesy of CCTV Labs, www.cctvlabs.com

Figure 61: Examples of laboratory black and white and color camera resolution charts.

As horizontal resolution increases, finer detail can be observed in the image. To determine camera resolution, a camera is positioned so that it views the full chart with no background visible (see Figure 62). When positioned correctly, the little arrows at the edge of the resolution chart align with the edge of the monitor viewing area. Observing the video monitor, at some point along the vertical converging black and white lines, the lines merge. At this point, four distinct black and white transitions can no longer be discerned. The point where the lines barely appear separate is the horizontal resolution. The resolution in HTVL is read from the numbers on the chart. For example, if four distinct lines are no longer visible at a point a quarter of the way between the 500- and 600-line calibration points, the camera would have a 525-line horizontal resolution. Figure 63 shows examples of field resolution charts.

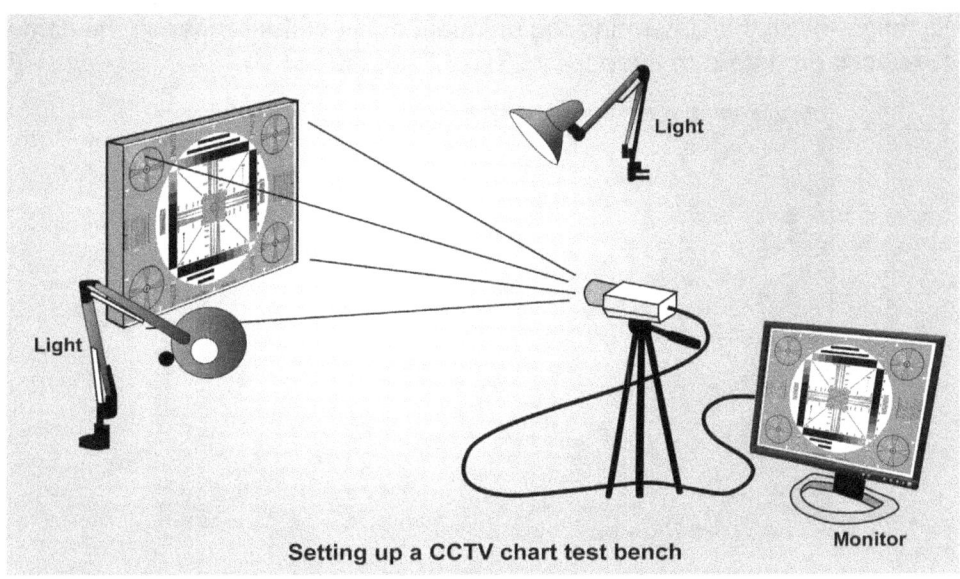

Figure 62: Example of a resolution chart test setup.

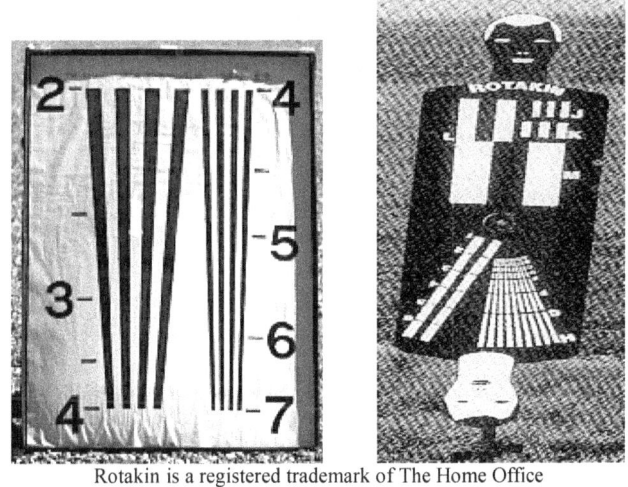

Rotakin is a registered trademark of The Home Office

Figure 63: Examples of field resolution charts (Rotakin at right).

As a point of reference, for real-time assessment, the horizontal resolution of a good quality video camera is about 800 pixels. Historically, 1 percent, or 8 pixels per foot of horizontal object resolution has been used as a basis for the resolution needed for intruder classification. This indicates that the maximum horizontal field-of-view should be limited to 30.5 meters (100 feet). It is important to note that object resolution specification in pixels per foot must encompass all of the camera-to-monitor elements (e.g., camera, lens, transmission system, video recorder, and monitor) to determine the video assessment system's end-to-end resolution.

For purposes of alarm assessment, three levels of resolution need to be considered:

(1) Detection—Ability to detect that an object is present in the camera's field of view; however, exact identification of the object may not be possible (see Figure 64).

(2) Classification—Ability to determine whether the object in the camera's field of view is human or nonhuman (see Figure 65).

(3) Identification—Ability to determine the unique identity of the human in the camera's field of view based on details of appearance (see Figure 66).

Figure 64: Example of video image showing detection.

Figure 65: Example of video image showing classification.

Figure 66: Example of video image showing identification.

These three levels of resolution are dependent on the camera's resolution and the size and proximity of the object to the camera. For example, depending on camera resolution, object size and object distance from the camera, it may be possible to identify a particular person, classify an object such as a large dog, or detect an object the size of a small animal. Consideration of the object or assessment target is critical in determining camera placement and the number of cameras required. In an exterior perimeter situation, a security system operator may need to classify a person crawling slowly through an assessment zone at night. The person crawling could be close to the camera or at the camera's far field of view.

The distance from the camera to the assessed area's far field of view is limited by a design criterion requiring that a specified number of lines of resolution (and thus a number of horizontal imager pixels) are occupied by a certain sized target. The amount of resolution required is determined by the level of assessment needed at a particular camera location. When assessment requirements change from intrusion detection to classification to intruder identification, the number of lines or pixels that a target must occupy on a video monitor screen increases.

Table 1 shows the number of horizontal pixels that are required for the three levels of assessment. In high-resolution camera imagers, the relationship between pixels and HTVL is 0.75. For example, it is recommended for assessment classification of a human intruder that there be eight horizontal pixels on a 1-foot target. From above, eight pixels equal six HTVL on a 1-foot target. With eight pixels or six HTVL on a 1-foot target at the camera's far field of view, the resultant picture quality allows the security system operator to be able to recognize and discriminate between human and animal.

Table 1: Assessment Type and Required Pixels

Assessment Type	Pixels
Detection	2 to 3
Classification	6 to 9
Identification	10 to 16

Distance from the camera, camera resolution, lighting, and other video assessment system performance characteristics contribute to how easily and quickly an operator can assess a situation. For exterior perimeter applications, having a camera resolution that allows an operator to classify a target is most likely sufficient to differentiate between an adversary attack (crawling persons) or a nuisance alarm such as a rabbit or crocodile. At the other extreme, for some interior applications, it may be desirable to uniquely identify a human in the camera's field of view.

6.2.2 Types of Cameras Available

This section provides an overview of camera technologies available for use in video assessment applications. Topics cover generic camera classifications such as color, black and white, and thermal, as well as technical attributes of some cameras such as auto shutter, auto iris, and integrating types.

6.2.2.1   Color and Black and White Cameras

While a color camera enhances daylight scenes by providing additional color information about the target being assessed, nighttime use of color cameras, particularly when a scene is illuminated with high-pressure sodium (HPS) lamps, can be problematic. Security perimeter field tests of color cameras with HPS lights have not been favorable. At night, HPS lamps cause a gold-orange color to appear on monitor images; for example, a perimeter floor composed of light-grey, rounded river stones appears orange. At maximum distances, it is difficult to distinguish an intruder's exact clothing color under the varying lighting conditions encountered in locations illuminated with HPS lamps. To the human eye, HPS illumination can distort naturally occurring color cues.

When comparing color and black and white cameras with the same type of camera imager, color camera resolution is about 18 percent less than that of an equivalent black and white camera imager. The black and white camera provides a brighter, sharper, higher-contrast image at night because of its higher resolution and operation only in the grey scale. More discussion is contained in the "Day/Night Cameras" section.

6.2.2.2   Day/Night Cameras

Some cameras produce color images during the day and black and white images at night, taking advantage of the best features of both. The camera has a sensor that measures the ambient light level and controls switching from day to night mode. On some day/night cameras, the ambient light level at which the switchover occurs can be adjusted. The camera electronics monitor the output video level and switches from color to black and white when the scene illumination level is less than a predetermined level. When transitioning from day to night mode (color to black and white mode), the camera mechanically removes a near-infrared filter. With the camera in the black and white mode, it then responds to near-infrared illumination blocked by a filter during the day. Removing the filter also improves picture brightness. This increases the amount of scene illumination reaching the camera imager. The nighttime black and white image has approximately 18 percent more resolution than the daytime color image.

6.2.2.3   Electronic Shutter Cameras

The pixels in camera imagers can be likened to electronic capacitors. Instead of accumulating electrical charge as capacitors do, camera imager pixels accumulate light charges over an active charging time and then send that charge information to be processed into a complex electronic signal for the monitor to display. Some cameras adjust the amount of time that the camera imager is allowed to be exposed to a scene for producing a frame of video. These cameras have electronics that automatically adjust the amount of imager exposure time as a function of scene illumination level. Exposure time is very short during sunny environments and is lengthened in darker nighttime environments.

6.2.2.4   Integrating Cameras

When scene illumination is too low and grainy pictures occur, some camera models can increase the time over which camera imagers are exposed to the scene. Known as integrating cameras, the length of time that the imager is allowed to be exposed to incoming illumination is increased to make the image brighter and improve scene contrast. These cameras slow the shutter speed to greater than 1/30th of a second when viewing very dark scenes to obtain enough scene illumination to produce a picture with sufficient contrast for assessment purposes. Some cameras can slow the shutter speed to as long as 4 seconds, but that length of time between frames is too long for adequate assessment. Integrating light for longer periods of time causes moving images to become blurred. For good assessment, the shutter speed should not exceed ¼ of a second. An intruder running a 4-minute mile will traverse 1.7 meters or 5.5 feet in ¼ of a second.

6.2.2.5   Intensified Low-Light Cameras

In a camera that has an intensifier component, photons are linearly accelerated and bombard a luminescent (green) screen. A color camera is focused on the green screen to create the displayed video signal. The intensifier responds to near-infrared illumination from stars, the moon, and artificial lighting. These passive cameras are only light receivers and do not emit near-infrared energy to illuminate the scene. The intensifier requires replacement and maintenance and has a limited life on the order of 2,500 to 3,500 hours of operation. Bright light sources in the scene can create washout smears or streaking in the camera image.

6.2.2.6   Thermal Imager Cameras

The thermal imager camera converts thermal radiance to a video signal. Its camera video output can be a black and white image showing the temperatures of objects as shades of grey or in a gradation of colors calibrated to temperature bands. Thermal camera imagers have resistive element pixels that respond to thermal/infrared energy in the 3 to 5 or 7 to 14 micron waveband emitted from warm-bodied sources. A thermal camera is a night vision device that responds to differences in temperatures against a background temperature reference. A thermal camera is a passive device and requires no illumination to produce a video image. The picture displayed is based on the difference in temperature of objects in the scene. Cooled thermal imager cameras typically use nitrogen cooling loop Sterling pumps. The advantage of cooled thermal imagers is that they have higher sensitivity to small changes in temperature. The Sterling pumps, however, require replacement at 10,000- to 15,000-hour intervals.

6.2.2.7   Pan-Tilt-Zoom Camera

The pan-tilt-zoom (PTZ) camera provides alarm station operators a level of versatility in performing surveillance tasks not possible with the use of fixed cameras. With a PTZ camera, the operator can rotate the camera over a 360-degree field of view, move the camera up and down, and zoom in to more closely observe activities of interest. PTZ cameras come in a range of mechanical configurations, including an exterior mount, dome-type camera; an exterior-mount camera configuration with fixed cameras mounted on an integrated pan-tilt mechanism (lenses can be fixed or zoom); an enclosed, ceiling-mount dome indoor camera, and a nonenclosed indoor camera.

6.2.3 Concepts for Cameras

6.2.3.1   Imager Format

The camera's format designator is based on the size or *format* of the image device. The camera format is related to the size of the photosensitive surface and is a measure of the diagonal of the scanned rectangular area.

As shown in Table 2, common imager formats today are in inches: 2/3, 1/2, 1/3, 1/4, 1/6, and 1/8 inch. The inch measurement is a carryover from tube image cameras and relates to tube diameter, not the actual measurement of the diagonal of the imager. Early analog video cameras had 1-inch-diameter tube-type imagers. A 1-inch tube-type imager had a 16-millimeter (mm) (0.63-inch) active area. When solid-state imagers were manufactured, the sizing convention continued, so that an imager with a 16-mm diameter and 12.8-mm horizontal dimension was referred to as a 1-inch imager. Subsequently, as smaller imagers were produced, the imager format size relationship continued with the existing solid-state sizing convention.

The ratio of the imager width to height is called the "aspect ratio." Standard video camera imagers have a 4:3 horizontal to vertical size (aspect) ratio that represents two sides of a 3-4-5 triangle. The vertical dimension is 0.75 times the horizontal dimension and the diagonal is 1.25 times the horizontal dimension. Imagers for high-definition television (HDTV) cameras have a 16:9 horizontal to vertical size ratio where the vertical dimension is 0.5625 times the horizontal dimension and the diagonal is 1.1475 times the horizontal dimension.

Table 2 Imager Formats, Converting Inches to Millimeters

Stated Format	Active Area	Imager Size (H x W)
1 inch =	16 mm diameter =	9.6 mm x 12.8 mm
2/3 inch =	11 mm diameter =	6.6 mm x 8.8 mm
1/2 inch =	8 mm diameter =	4.8 mm x 6.4 mm
1/3 inch =	6 mm diameter =	3.6 mm x 4.8 mm
1/4 inch =	4 mm diameter =	2.4 mm x 3.2 mm
1/6 inch =	3 mm diameter =	1.8 mm x 2.4 mm
1/8 inch =	2 mm diameter =	1.2 mm x 1.6 mm

6.2.3.2   Scene Contrast

Video image contrast relates to the difference between the white and black video signal levels in a picture image. High contrast indicates a large difference between the white and black levels, while low contrast indicates a small difference between the white and black levels. A low-contrast video image appears as shades of the same color hue. Figure 67 shows images of high- and low-contrast video scenes.

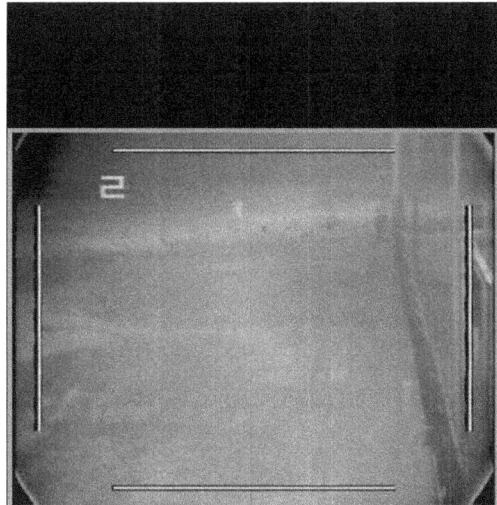

Figure 67: Images of high-contrast (left) and low-contrast (right) scenes.

6.2.3.3 Sensitivity

The sensitivity of a video camera can be defined as the minimum amount of illumination required to produce a specified output signal. The following factors are involved in producing a video signal:

- Illumination level of the scene

- Color spectral distribution of the illuminating source

- Object reflectance

- Total scene reflectance

- Camera lens aperture

- Camera lens transmittance

- Spectral response of the camera imager

- Video amplifier gain, bandwidth, and signal-to-noise ratio

- Electronic processing circuitry

Camera sensitivity may also be expressed as scene illumination or the level of imager illumination that will produce a minimally acceptable video image. Normally, camera sensitivity is stated in units of lux or foot-candles. For practical purposes, experts in the field indicate camera sensitivity to be the illumination level that produces a 60-percent reduction in camera output. While usable camera video can be observed at levels below a 40-percent signal level, further reduction beyond 40 percent increases the amount of video signal noise and picture graininess. Below (Figure 68) is an excerpt from a camera specification page showing a sensitivity specification.

6-13

Sensitivity		
	Full Spectrum	With IR Filter
Full Video, No AGC	0.039 fc (0.39 lux)	0.15 fc (1.5 lux)
80% Video, AGC On	0.001 fc (0.01 lux)	0.006 fc (0.06 lux)
30% Video, AGC On	0.0002 fc (0.002 lux)	0.0009 fc (0.009 lux)

Figure 68: Camera sensitivity at light level.

6.2.3.4 Signal-to-Noise Ratio

Camera signal-to-noise ratio (SNR) is the strength of video signal above the electronic noise level created by the camera's electronic signal processing circuits. SNR is expressed in decibels (dB). The higher the SNR, the better the rating. With low SNR, the picture can appear grainy or snowy and sparkles of color may be noticeable. Noisy video produces poor-quality pictures. A good camera SNR is 50 dB or higher.

6.2.3.5 Automatic Gain Control

Automatic gain control is an electronic circuit in the camera that maintains a preset video output signal by increasing electronic amplification of the video signal for low illumination scene conditions. A primary method to control the amount of light reaching the camera's imager is the motor-operated automatic iris lens system. The camera's output video signal amplitude is monitored either externally by a circuit in the lens assembly or internally within the camera, and an iris-control drive signal is created to maintain the amplitude at some predetermined output level. As the video level decreases (meaning that there is not enough illumination reaching the camera's imager), the auto-iris lens is driven to increase the size of the aperture. At some point, however, the iris is driven fully open, and further decreases in light level result in a reduced amplitude video signal. At this point, the electronic amplification is increased. As the light level is reduced even further, the improvement caused by electronic signal amplification is exhausted, and a very grainy picture is presented. To prevent grainy pictures, some cameras have a cutoff circuit that causes the camera to produce no output when picture graininess exceeds a certain threshold.

6.2.4 Characteristics and Applications for Cameras

Several camera characteristics enhance a camera's ability to provide reliable assessment images over varying environmental conditions, as listed below:

- High sensitivity can provide the brightest and highest contrast video image under widely varying lighting conditions.

- Automatic gain control, auto-iris control, or auto-shutter control circuits maintain video image quality over a range of day and night lighting conditions. These features maintain good picture quality while bright lights such as vehicle headlights are in the scene's field of view. The imager's persistence is short enough to preclude image smearing during movement, and the image does not have a washout streak above and below a bright light in the camera's field of view.

- The camera has a history of reliability, durability, and resistance to environmental weather effects.

6.2.4.1 Exterior Cameras

Exterior cameras should be mounted on stable towers and mounts so that motion or movement in the wind is avoided. A wire frame tri-pole tower (instead of a wooden pole) is unaffected by varying weather conditions.

Exterior cameras should be mounted at heights of 7.6 to 9.1 meters (25 to 30 feet) above the assessment area surface. The cameras should be tilted down to view the entire assessment area. On flat perimeter surfaces, camera downward tilting on the order of 2 to 5 degrees is recommended. With the cameras tilted down, the horizon is not in the camera's field of view, and glare at sunrise and sunset is reduced. In addition, particularly if the camera's enclosure is covered with a sunshield, snow and ice impingement on the camera enclosure's front cover glass is reduced. Exterior lighting luminaires should be at least 3 meters (10 feet) higher than the cameras. For example, if the cameras are mounted on top of 9.1 meter (30 foot) towers, lighting luminaires should be affixed to poles that place them at least 12.2 meters (40 feet) above the ground. All camera systems, associated power supplies, and connections should be protected from unauthorized external manipulation such as tampering.

6.2.4.2 Camera Enclosures

Exterior cameras require environmental enclosures to protect cameras from outdoor temperature extremes, dust, dirt, humidity, wind, rain, and snow. The enclosures must be large enough to contain the camera, lens assembly, power supply, and, possibly, camera communication modules. Two types of environmental enclosures are described below:

- An integral environmental enclosure is normally cylindrical in shape and has an O-ring sealed cover glass and rear cover. The enclosure is of rigid construction and can be pressurized with dry nitrogen and has integral front cover glass heaters. A sunshade protects the camera enclosure from high summer temperatures by shading the enclosure and providing a shade to cover the front cover glass of the enclosure.

- A sheet metal, formed metal, or fiberglass enclosure permits camera access through a clamshell-type cover (like the hood on a car). The cover is either hinged or removable. These enclosures cannot be pressurized. The large sealing surface and enclosure manufacturing techniques normally do not provide for a dust-tight seal. The primary advantage of the clamshell-type enclosure is for ease of accessibility for camera/lens replacement and for ease of lens focus adjustment.

Accessories for exterior enclosures include heaters, insulation material, fans, defrosters, and front cover glass washers and wipers. Washers and wipers tend to be high-maintenance items with the need for frequent washer solution and wiper blade replacement.

6.2.4.3 Pan-Tilt-Zoom Cameras

In addition to fixed alarm-assessment cameras, PTZ cameras have been installed in facilities as an added capability for tactical surveillance. These cameras require controller electronics and an operator joystick control to facilitate camera movements. While PTZ cameras are not recommended for alarm assessment, they are effective for monitoring temporary operations such as a construction crew working near a fence line. They can also be used as a means to facilitate a compensatory measure in the event of an assessment camera outage. A PTZ camera can be positioned to view the area covered by a defective fixed camera and joystick control locked-out for that camera.

6.2.5 Additional Considerations for Cameras

Camera capabilities can be inhibited by positional errors in camera placement, mismatches in expected and actual resolution, overt or covert tampering, environmental conditions, and overall system response time. The relationship between expected camera resolution and actual need is described in the resolution section above. For example, if target identification is the security objective and the chosen camera resolution can provide only target detection, then the camera and lens installed at that location are inappropriate for the application.

Covert video signal tampering can be accomplished in analog camera systems by tapping into video transmission cables and inserting a recorded scene or by switching video feeds to show video from the wrong zone. Overt tampering includes the following:

- Using a bright light or laser to blind the camera

- Covering the camera

- Shooting the camera lens or enclosure cover glass with a paint gun

- Cutting video transmission cables

- Destroying the camera to make it totally nonfunctional

Placing a camera in an unprotected area could lead to undetected camera tampering and attack on the camera's video transmission infrastructure. Further, the camera tower could also be used to enable egress across a perimeter, which would circumvent detection by ground-based sensors.

Changing environmental conditions such as rain, fog, or blowing snow can be the source of video assessment system weaknesses. These conditions can cause the loss of usable images, necessitating the implementation of contingency plans to provide an alternate form of alarm assessment, such as dispatching patrols to the area.

When designing an intrusion detection system, the sensored areas and the video assessment of those areas should be considered together so that the entire detection area can be assessed. The design should ensure that there are no locations in the sensor detection area where an intruder could hide to avoid camera assessment. Cameras, mounts, and towers are placed so that they do not interfere with or compromise sensor performance. Camera towers should not be placed so close to the sensor detection area or volume as to create nuisance alarms, decrease sensor sensitivity, or be used by the intruder as a climbing aid to bridge a sensor's detection zone. In multiple sensor configurations, the assessment camera should be able to view the combined sensored areas and volumes, as well as all of the hardware such as junction boxes or field data panels associated with the sensor. Sensor hardware should not be large enough to provide a convenient hiding place for intruders in video assessment areas.

Wooden poles are not recommended because they will dry out and twist over time. To compensate for twisting action of the wood, cameras must be repositioned from time to time to maintain the proper view of the area to be assessed.

Video assessment in shadows surrounded by bright sunlight creates significant problems because of the high scene dynamic range in brightness levels. In these circumstances, a compromise must be made to select the area of prime interest and to ignore the unresolved areas.

If **exterior cameras** are positioned so that the horizon is in the camera's field of view, it is possible (particularly for east- and west-facing cameras) for the rising or setting sun to be in the camera's field of view, blinding the camera and allowing an adversary to enter the perimeter without being adequately assessed. Similarly, **interior cameras** with lights in the camera's field of view can experience a glare or bright spots that will wash out usable images. In addition, a camera focused under one level of light and operated under a different light level or with a different lens mount or format will result in improper focus. Exterior cameras should be focused with the iris fully open at dusk to obtain optimum focus through the entire depth of field. Incorrect camera placement or lens selection can result in a horizontal field of view that is too narrow for the near field or not sufficient to see an individual or other target at the end of the assessment zone.

Outdoor cameras should be installed so that no light sources are in the camera's field of view. Direct light can cause image "blooming" or allow the auto-iris lens to close to reduce the amount of light allowed through the lens. Possible light sources include perimeter lighting, sky, exterior lighting on buildings, car headlights, and shiny objects that reflect light. Cameras should never be aimed so that the horizon and sky are in the camera's field of view. Because a camera is a light-averaging device, images of the ground surface (particularly at low sun angles such as sunrise and sunset) tend to be darker and therefore harder to assess. Considerable care must be taken because camera blinding from unexpected light sources is difficult to predict before installation and is one of the most frequent problems encountered after equipment is installed. Illumination sources in a camera's field of view may have to be shielded or reoriented to prevent the creation of a bright spot in the camera's field of view. Perimeter isolation zones adjacent to a roadway can also be problematic. Reflections from vehicle headlights and tail lights from a roadway surface can affect alarm assessment capability.

PTZ cameras should not be used for alarm assessment because of timing, reliability, and operational issues. PTZ cameras may compromise effective, timely alarm assessment. Use of PTZ cameras presents a high likelihood that they will be pointed at the wrong location or be zoomed so that part of the area to be assessed is not in the camera's field of view. Only fixed cameras with fixed or manually adjustable zoom lenses should be used for alarm assessment. This provides pre- and post-alarm video of the entire assessed zone. PTZ cameras should be used only for tactical surveillance as an added feature to augment alarm assessment cameras or for backup in the event of primary camera failure.

6.2.6 Criteria for Cameras

The following considerations in camera selection are listed in order of priority:

(1)   The main consideration in camera selection is the **sensitivity** required for a full video output signal for the lighting environment in the area to be assessed. To meet the desired or required performance capabilities, the sensitivity must match the lighting design goals, regardless of the imager.

(2) The **resolution of the imager** is next in importance because it will determine the number of cameras required for a given straight-line perimeter section. The greater the resolution, the greater the spacing between cameras can be. The object resolution required should be determined before the camera is selected, but in practice, the desired object resolution may be slightly modified when the possible camera choices are limited.

(3) *Camera* **format** is an important consideration in the camera selection process. The requirements of nonstandard-sized focal length lenses should be carefully considered and evaluated before choosing a format.

(4) During the selection process, camera evaluation should consider the nighttime **lighting environment** expected at the site. Manufacturers' literature should not be the only basis for selecting a camera. The manufacturers' specifications or test conditions may not match the environment at a particular facility.

(5) Other considerations in the selection process should include **the difficulty in performing maintenance, the packaging of the camera for the intended environment, the manufacturer's maintenance support, and the documentation supporting the equipment.** Documentation should include operating, adjustment and maintenance procedures, theory of operation, block diagrams, schematics, and replacement parts lists. Serious consideration should be given to eliminating any manufacturer's product that does not include this documentation.

Significant factors that affect the system performance of exterior location video assessment include the following:

- Perimeter layout

- Lighting layout

- Weather effects (fog, heavy snow, etc.)

- Surface conditions (flat with no hills for hiding, evenness of scene reflectiveness, rain, drainage)

6.2.7 Testing Cameras

Section 6.5 "Video Assessment System Testing" within this document contains testing information that should be approached from a systems point of view. This testing section addresses several kinds of testing, including acceptance, operational, and performance testing, integrating cameras, DVRs, lighting, and system elements.

6.3 Digital Video Recorders

6.3.1 Principles of Operation

Over the past 10 years, DVRs have almost totally replaced video cassette recorders (VCRs) for use in security video assessment applications and have significantly improved the process and quality of that video. DVRs record video onto arrays of hard drives and so do not require the changing or rewinding of tapes. The most typical DVRs on the market today, depending on the features purchased, can do the following:

- Record the images obtained from 1 to 16 cameras simultaneously

- Be set to record only when motion occurs in the camera scene, thus saving storage space

- Instantaneously access recorded video from a particular time period

- Store huge amounts of recorded video for weeks or even months

- Adjust the number of frames per second of video to store from each camera, as well as the resolution of that stored image

Modern DVRs are computer-based devices and can be controlled by an interactive system with the sensor alarm monitoring portion of a security system. Upon sensor alarm notification, the DVR can be directed to play back pre- and post-alarm video from the camera assessing the area covered by the alarming sensor.

Video image recording is a process using a personal computer (PC) to capture a video stream and store that video information to a network of computer hard drive memories. It is then possible to play back video streams from the hard drive for operator display. The components of a video image recording system are a PC with video recording card, video recording software for managing the storage and playback of the video images, and a computer monitor for displaying the video images. The video management software allows the computer to display incoming video on a monitor, record images, store incoming images to the hard drive, and play back stored images from several cameras on the monitor simultaneously.

Incoming camera analog video signals are separated into their main components (luminance and chrominance), then converted to a digital format so that they can be stored on a computer hard drive. **Luminance** contains the black and white portion of the video signal, and **chrominance** contains the color portion. Digital filters are applied to each pixel (picture element) of the video image to ensure that every single pixel of video image is represented in digital format with the highest accuracy. Once the analog video signal has been converted to a digital signal, a certain amount of signal noise (white noise and other visual imperfections) results from the conversion process that needs to be removed before the next step of video processing, called **compression**. Noise reduction software algorithms clean up the digital video information to provide better video quality, improve video encoder compression efficiency, and improve video content of the stored images.

To maximize the amount of video images that can be stored on a particular DVR system, **video compression** uses a coder decoder (codec) to compress video content into a reduced size format using an encoding scheme to fit efficiently onto a hard drive memory. For example, without compression, a 2-hour movie would need to be stored on 30 DVDs rather than just one. Compression encoders include Motion Picture Experts Group (MPEG format), Joint Photographic Experts Group (JPEG format), Motion JPEG (M-JPEG format), H.2.64, and Wavelet. The H.2.64 compression has its roots in coding for videoconferencing and video telephony, as well as streaming broadcast, file download, and media storage and playback. Wavelet algorithms for video compression originated in Europe. DVR manufacturers have either adopted a compression scheme derived from one of the types described above or have created their own proprietary compression algorithm by combining portions of standard compression codecs. DVR manufacturers more often use MPEG encoder compression because of the efficiency of video compression and the quality of the reconstructed playback image. Compression encoders analyze the video and decide which pieces of video information can be eliminated because they do not contain important visual content or contain redundant information, such as an image background that is all the same color.

Once the video is stored in computer hard drive memory, retrieving images for playback requires that the video images be processed by several software algorithms before display on a computer monitor. The stored images must be uncompressed and reformatted to produce a series of pictures that, taken in timed sequence, replicates the initial analog images from which the stored video was obtained. Reconstituted images can contain rough or harsh edges, so a smoothing algorithm is applied to the edges of objects to make them appear more natural. Finally, a scaling algorithm is applied to adjust the picture to the size of the screen and clean up the edges of the picture. Scaling is also applied when image size is adjusted to display images simultaneously from several cameras on the same monitor screen or change the shape of the image displayed from its initial image size.

The standard frame rate for analog cameras operating to Electronics Industry Association standards is 30 frames per second. The maximum DVR video recording rate is specified in frames per second (FPS). Most DVRs can record video at rates of at least 30 FPS. However, many cannot record simultaneously at that frame rate on all channels (each channel serves one camera). The maximum recording rate in FPS must be divided among all the cameras served by a DVR unit, though this does not have to be an equal number of frames for each camera. If a 16-channel DVR has a maximum recording rate of 240 FPS, on average, if all channels are being recorded at the same frame rate, a maximum of 15 frames per channel can be allocated per channel. Because DVRs are computer based, the number of frames to be captured from each camera can be customized based on the camera, the time of day, or the type of received alarm that initiates the recording.

6.3.2 Types of Digital Video Recorders Available

DVR technology is evolving at the same rate as general computer technology. Capabilities and features of commercial DVR systems change radically every 12 to 18 months.

Some DVRs can control as many as 32 to 64 cameras each, though this feature is not necessarily advantageous if the system goes down; multiple DVR systems controlling only 16 cameras each may be preferable, as a significant number of a facility's cameras will not be rendered unavailable if one system becomes unavailable.

The number of video frames a particular DVR can record at any one time is also increasing dramatically. Minimum systems will record 30 FPS; newer systems can record as many as 240 or more FPS.

6.3.3 Minimum Configurations and Installation and Setup Requirements for Digital Video Recorders

A DVR should contain a reasonably fast microprocessor, which at a minimum, should possess the following characteristics:

- At least 1 gigahertz clock speed or higher

- At least 256 megabytes (MB) of RAM

- At least a 5,400-rpm, 300-gigabyte hard drive

- A video recording card with at least 64 MB of onboard memory

- Special software dedicated to simultaneously processing multiple streams of video and responding to control commands from a security alarm control and display system

The system should have at least a 17-inch monitor (preferably larger where possible) with at least a 1,024 × 768 pixel 32-bit color display. A 10/100 MB network connection is required for streaming digital video from the video image recording computer to the operator's console. Existing analog cameras can connect directly to DVR analog inputs. Digital cameras can connect to recorders called network video recorders using Ethernet networks. A DVR can be implemented as a group of stand-alone boxes, or it can be integrated into a desktop computer-type configuration.

Most DVR systems are considered to be user friendly, employing simple VCR-type control commands such as play, stop, fast-forward search, fast-reverse search, and enhanced digital control commands to play back-tabbed video associated with a specific alarm event. Because the DVR is computer based, many setup and control parameters can be customized for each application.

6.3.3.1   Raw Camera Images versus Recorded Images

To determine if the configuration of a DVR (or even the DVR itself) is adequate for purposes of assessment video playback, the output of the raw camera image should be compared to the same image played back from the DVR. Viewing the output of a DVR while recording is taking place is **not** a definitive means for making a comparison. Most DVRs have a loop-through output that takes the incoming camera input video and connects it to the video output connector. What is recorded and subsequently (and instantaneously) retrieved from the hard disks may have a significantly reduced resolution from the raw camera image. Therefore, it is important that when conducting resolution tests, actual playback images from the hard disk be compared to the raw camera images.

6.3.3.2   Digital Transport

A DVR can be used to copy video clips and images to portable memory devices for transport from the alarm station and for evidentiary purposes. Networked DVRs can be set up to send an e-mail with alarm video to response force personnel or offsite personnel for immediate response.

6.3.3.3   Memory Management

A DVR's memory management can be configured to overwrite previously saved video when the maximum memory storage limit is reached or when the oldest video has exceeded its maximum storage time limit. DVR output display configurations allow for single or multiple camera images to be viewed as separate windows or panes on digital display monitors. Monitor views can be configured to simultaneously display live and recorded alarm video.

6.3.3.4 Video Motion Detection

Video motion detection (VMD), a feature available in most DVRs, allows the DVR to automatically store video when a movement within a camera's scene is detected. It can be set up to allow secure access to recordings using the Internet and secure communications and access controls. (See Figure 69.)

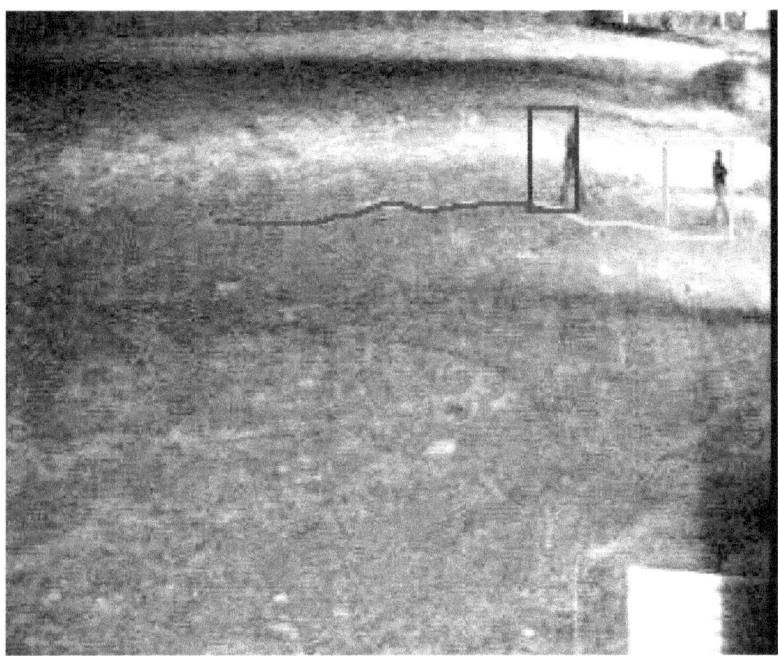

Figure 69: Video motion detection image showing intruder tracking.

The video storage capacity of a DVR system is determined by the following factors:

- Amount of hard disk drive storage space

- Number of simultaneous video channels being recorded (i.e., the number of cameras)

- Number of frames captured per second (frame rate) for each camera

- Video image resolution (pixels per frame)

- Quality of video recorded (compression)

The file size of recorded images can be controlled by reducing the captured resolution. DVR resolution is represented as a pixel count or in a video format term called "common intermediate format" (CIF). CIF is a standard video format used to define resolution in terms of pixel count. Table 3 shows format in terms of CIF, pixel resolution of that format, and the size of an uncompressed video image with that format.

6-22

Table 3: *Image Sizes for Common Intermediate Format Image Sizes*

Format	Resolution (Pixels)	Uncompressed Image Size (kB)
CIF	352 × 288	250
QCIF	176 × 144	63
4CIF	704 × 576	1,000
16CIF	1408 × 1152	4,000

The size of the video files stored on the hard drives can be reduced by increasing the degree to which each captured image is compressed. A good camera may provide a high-resolution image, but the captured image may be compressed to save storage space. To view a recorded, compressed image, the image is decompressed, but when viewing the decompressed image, image detail can suffer. Overall image quality is a function of camera resolution, captured image resolution, and the amount of compression applied to the original digital image.

6.3.4 Testing Digital Video Recorders

The last portions of this section contain testing information that should be approached from a systems point of view. The testing section addresses several kinds of testing, including acceptance, operational and performance testing, integrating cameras, DVRs, lighting, and system elements.

6.3.5 Digital Video Recorder Maintenance

Maintenance for a DVR system is minimal. The manufacturer's suggested maintenance schedule should be followed.

6.4 Lighting

6.4.1 Lighting for Video Assessment

6.4.1.1 *Principles of Operation*

Lighting for alarm assessment allows security personnel to maintain visual assessment capability during darkness. When security lighting provisions are less than optimal, additional security posts, patrols, night-vision devices, or other provisions are necessary for acceptable alarm assessment.

Security lighting should be used in vital areas and along perimeter fences when the situation dictates that the area or fence must be under continuous or periodic observation during nighttime hours.

Lighting when properly used in conjunction with video cameras may reduce the number of security personnel needed for alarm assessment. It may also enhance the personal protection of security personnel by reducing the possibilities of concealment and attack from a determined intruder.

Security lighting is desirable for those sensitive areas or structures within and at the perimeter that are under constant video observation. Such areas or structures include perimeter isolation zones, vital buildings, storage areas, and vulnerable control points of communication, power supply, and utility infrastructure systems. In interior areas where night operations are conducted, adequate lighting facilitates the detection of unauthorized persons approaching or attempting malicious acts within the area. Security lighting also has considerable value as a deterrent to intruders and may make the job of the adversary more difficult. Lighting is an essential element of an integrated physical security program.

A secure emergency power source (such as an uninterruptible power supply (UPS)) and power distribution system for the facility should be installed to provide power source redundancy for critical security lighting, as well as for security detection and assessment, control, and monitoring equipment. If primary power is temporarily lost as the result of power system outages or hostile activity, an emergency power supply enables critical security equipment (e.g. detection, assessment, illumination, control, and monitoring assets) to remain operable thus maintaining the integrity of the physical protection system employed at the facility. Emergency power sources should be available immediately without functional interruption for critical electrical loads and should be secured against direct and indirect attack as well as sabotage.

Perimeter lighting needs should be based on the threat, site conditions along the perimeter, video assessment capabilities, and available security personnel. Security lighting should be designed and operated to facilitate the detection of intruders approaching or attempting to gain entry into protected areas and to discourage unauthorized entry.

An effective lighting design is of paramount importance to the proper functioning of alarm assessment systems. Just as humans need good lighting to see, most security video cameras require appropriate lighting to allow for efficient assessment of the area when natural light is not adequate or available.

Intuitively, it would seem that an organization could order poles and an exterior luminaire to install at the top of each pole, run power lines, and then set the poles and light fixtures evenly spaced along the areas to be assessed. It might also seem that brighter lamps would require fewer poles and luminaires. However, that concept is not the optimal approach from a camera assessment perspective. To compare how lighting affects assessment, note the differences in the evenness of illumination between Figure 70 and Figure 71. Figure 70 has one very bright spot that affects camera performance and dark spots where an adversary might hide. In the event of an alarm condition, the scene depicted in Figure 71 would allow for a more efficient video assessment to determine the cause of the alarm so that security personnel can be appropriately dispatched.

Figure 70: Example of lighting with hot spots, dark areas, and dirt ground cover.

Figure 71: Assessment visibility improved with even lighting and a regular ground surface material.

If a facility is using thermal imaging cameras as the means for providing alarm assessment at night, illumination in the specific areas covered by these cameras is not as critical; however, it may still be required to provide a means for the security force to identify and engage potential adversaries using a force continuum. If the potential exists for the security force to use deadly force for the protection of the facility, illumination is a critical component that enables security force members to properly identify and accurately engage the adversarial force after detection and assessment are complete.

6.4.2 Installation Criteria

Security lighting usually requires less illumination than normal task lighting, except for personnel identification and vehicle inspection at an entry control point. Each area of a facility presents its own unique set of considerations based on physical layout, security requirements, terrain, and environmental and climatic conditions. Information that is available from lighting equipment manufacturers and vendors of lighting analysis software to assist in designing a lighting system includes the following:

- Descriptions, characteristics, and specifications of luminaires and lamps

- Luminaire lighting patterns

- Installation layouts showing height and spacing of luminaires to achieve light levels Desired

- Software to produce computer-generated plots of illumination levels and lighting uniformity in a particular zone and summary statistics of the illumination profile

  - Average illumination level
  - Light-to-dark ratio
  - Maximum and minimum illumination

In planning a security lighting system, the site designer should consider the following:

- The cost of replacing lamps and cleaning fixtures, as well as the cost of providing the required equipment (such as ladders and bucket trucks) to perform this maintenance

- Provision of automatic transfer or manual-override capability during a loss of primary power

- Photoelectric controls for automatic control of lights during nighttime hours

- Effects of local weather conditions on lighting systems

- Fluctuating or erratic voltages in the primary power source

- Grounding requirements

- Provisions for rapid lamp replacement and luminaire cleaning

- Special lighting requirements for critical areas (such as protected area perimeters), lighting provided with the means to remain operable without interruption during the loss of primary power. Any amount of time without adequate lighting in a critical area may be unacceptable. Therefore, these areas generally have emergency power (such as uninterrupted power supplies consisting of batteries and diesel engine generators) in case of primary power loss.

- Continuous operation of security lighting systems during hours of darkness

- Security lighting system configured so that the failure of one or more lights will not affect the operation of the remaining lights

- Restrike time (the time required before a light will function properly after a brief power interruption)

- Color spectrum of bulbs

6.4.3 Principles of Security Lighting

Security lighting enables security personnel to observe activities using alarm assessment and surveillance cameras around or inside a facility while minimizing their physical presence throughout the facility. Having adequate illumination levels at all approaches to a facility does not necessarily discourage unauthorized entry. However, adequate lighting improves the ability of security personnel to visually assess intrusion alarms with the use of video cameras and intervene in the event of an unauthorized access attempt. Lighting is implemented with other security measures, such as intrusion detection sensors, video assessment equipment, and alarm control and display systems as part of an integrated facility security system.

Optimum security lighting is achieved by adequate, even light in perimeter isolation zones. Additionally, the use of deterrent lighting (i.e., glaring lights directed away from the fenced perimeter) can augment security perimeter lighting as a psychological deterrent to intruder ingress. In addition to seeing long distances, security personnel must be able to see low contrasts, such as indistinct outlines of silhouettes and must be able to detect an intruder who may be exposed to view for only a few seconds. Higher levels of illumination, such as that provided by deterrent lighting systems, if properly implemented, can improve these assessment abilities.

Contrast between an intruder and the background should be an important consideration when planning for security lighting. With predominantly dark surfaces, more light is needed to produce the reflective brightness required for camera assessment than would be necessary if neutral gray backgrounds and ground cover are used. When the same amount of light falls on an object and its background, the observer must depend on the contrast of light reflected from each to determine intruder location. The ability to differentiate between background and objects of interest when contrast is poor can be improved significantly by adjusting the illumination level or the lighting location. (See Figure 72.)

Figure 72: At left, poor contrast created by outdoor lighting; at right, better contrast created by outdoor lighting.

The observer primarily sees an outline or a silhouette when the intruder is darker than the background. Using light-color finishes on the lower parts of buildings and structures may expose an intruder who depends on dark clothing and darkening the face and hands. Placing stripes on walls has also been used effectively as patterns that can provide recognizable breaks in outlines or silhouettes.

Security lighting should be practical and effective. To be effective, it must discourage or deter intruder access attempts and, if access is attempted, it should facilitate detection and assessment. It is important that security lighting be designed to provide sufficient illumination so that security personnel can effectively observe an intruder with the naked eye but not be so bright as to create a glare that may temporarily blind the security personnel.

The eyes of security force response teams dispatched from building interior locations will not be night-adapted before they respond to an intrusion alarm at an exterior location where light levels are below 0.5 f-c (5 lux). Full sensitivity change (adapting to a significantly lower light level) may take up to 30 minutes, making proper illumination in exterior areas critical to the effectiveness of the responding security force.

6.4.4 Types of Lighting

The type of lighting system used depends on the installation's overall security requirements. Four types of lighting approaches can be used for security lighting systems: continuous, standby, movable (portable), and emergency.

6.4.4.1 Continuous Lighting

Continuous lighting is the most common type of security lighting used. As shown in Figure 73, it consists of a series of fixed lights arranged to continuously illuminate a given area during darkness with overlapping cones of light. Two primary methods of using continuous lighting are controlled projection lighting and deterrent glare projection lighting.

Figure 73: An example of a perimeter with controlled lighting.

6.4.4.1.1 Controlled Lighting

Controlled lighting is optimum when the limits of the width of the lighted strip are directed to illuminate the inside of a perimeter isolation zone with minimal illumination of the areas inside the perimeter. Having too much illumination on the protected side of the perimeter may illuminate or silhouette security personnel, giving the intruder an unnecessary visual advantage.

6.4.4.1.2 Deterrent Glare Projection Lighting

Deterrent glare projection security lighting (such as spotlights) is used when the glare of lights directed across the surrounding territory will not interfere with adjacent operations or be politically unacceptable to neighbors. These types of lights are a strong deterrent to a potential intruder because they make it difficult for a would-be intruder to see inside the area. Security personnel are protected by being in comparative darkness and being able to observe intruders who are at a distance outside the perimeter.

6.4.4.2   Standby Lighting

Standby lighting has a layout similar to continuous lighting. However, the luminaires are not continuously lit during nighttime hours but are either automatically or manually initiated when suspicious activity is detected or suspected by security personnel or the intrusion sensor system.

6.4.4.3   Movable Lighting

Movable lighting consists of manually operated, movable integrated luminaire and generator assemblies that may be operated during hours of darkness or as needed. This type of system is normally used to supplement continuous or standby lighting.

6.4.4.4   Emergency Lighting

Emergency lighting is a system of lighting that may duplicate any or all of the above systems. Its use is limited to times of power failure or other emergencies that render the normal system inoperative. It depends on an alternative power source such as installed or portable generators or battery-powered UPS.

6.4.5 Lighting Definitions

Illuminance, defined as the intensity of illumination incident on a surface area, is stated in terms of "foot-candles" or "lux," which are not equivalent measurements. One foot-candle (f-c) is defined as the intensity of illumination on a 1-square-foot surface from a 1-lumen source of light located 1 foot away. One lux is the international system of units (SI) standard unit (metric) for illuminance, defined as the illumination on a 1-square-meter surface from a 1-lumen source located 1 meter away. One f-c is equal to 10.76 lux. For ease of conversion, industry practice uses 10 as the conversion factor between unit systems (i.e., 1 f-c = 10 lux).

Illumination is either natural or artificial. Natural illumination emanates from the sun, moon, and stars. Artificial illumination is manmade and comes from an illumination source such as a luminaire.

Table 4 provides the approximate illumination levels for nine natural lighting conditions. As shown, the range of illumination from direct sunlight to starlight in an overcast sky spans about 9 orders of magnitude. However, cameras do not have the ability to compensate for that range of natural illumination. Therefore, luminaires must be used to illuminate assessment areas during nighttime hours. This reduces the dynamic range of camera compensation for daytime and nighttime lighting conditions to only 4 orders of magnitude.

Table 4: Relative Illumination Levels under Various Lighting Conditions

Direct Sunlight	3,200 to 13,000 f-c
Full Daylight (not direct sun)	1,000 to 2,500 f-c
Overcast	100 to 1,000 f-c
Sunrise or Sunset (clear day)	20 to 60 f-c
Twilight	0.1 to 10 f-c
Full Moon (clear)	0.01 to 0.1 f-c
Full Moon (overcast)	0.001 to 0.01 f-c
Starlight (clear)	0.0001 to 0.001 f-c
Starlight (overcast)	0.00001 to 0.0001 f-c

A scene becomes visible when illumination is reflected from surfaces and objects within that scene. **Reflectance**, the percentage of light reflected from a scene, depends on the angle of the light incident on the surface of a scene and on the texture and composition of the reflecting surface. Reflectance is determined by measuring illumination with a light meter's sensor facing down, divided by the illumination measurement with the light meter facing upward. (The measurement should be made such that neither the light meter nor the person making the measurements create a shadow on the sensor when face up or on the ground just below the light sensor when face down.) For natural illumination, the reflectance of various scenes is relatively independent of the angles of incidence and reflection. Table 5 lists some common surfaces and their approximate reflectance.

Table 5: Typical Reflectance for Various Common Surfaces

Surface	Reflectance (%)
Empty asphalt surface	7–10
Sandy soil, wet	15–20
Grass-covered area with trees	20–25
Gray rounded river rock	25–35
Red brick building	30–35
Sandy soil, dry	30–35
Unpainted concrete	35–40
Smooth surface aluminum	60–65
Snow-covered field	70–75

To ensure that a sufficient amount of the nighttime illumination is reflected back to the camera and that sufficient scene contrast exists for intruder detection, the ground surface of the assessment area should have a nominal reflectance of 25 to 35 percent when dry.

Illumination is typically measured using a light meter with the sensor face up at a specified distance above a horizontal ground plane. Normally, measurements are made at 6–12 inches above the ground. An area's average illumination is determined by taking several measurements of illumination at equally spaced locations in an illuminated area or zone. For example, measurements are taken in straight lines at 10-foot intervals in both the lengthwise and widthwise directions within a perimeter isolation zone beginning at one fence line and ending at the opposite fence line. Several measurements can be taken within an area or zone covered by a number of lamps and then averaged together. These are measurements of "horizontal scene illumination," or simply **scene illumination**.

Average scene illumination is the average amount of light illuminating an assessment area. The average scene illumination must be high enough to achieve adequate alarm video assessment, as well as visual assessment by security personnel. Tests have shown that an average scene illumination of 1.0 f-c onto a ground cover surface with 25 to 35 percent reflectivity provides sufficient light for camera and security personnel assessment purposes in clear environments.

The "**evenness**" (also referred to as "flatness") of scene illumination enhances the ability to assess intruder location. The scene at left in Figure 74 has a light-to-dark ratio of approximately 20-to-1 (20:1). The scene at right in Figure 74 has a light-to-dark ratio of approximately 4-to-1 (4:1). Tests have shown that lighting designs with at least a 6-to-1 (6:1) light-to-dark ratio at the end of bulb life provide sufficient illumination evenness for assessment purposes.

Figure 74: The scene at left has a high light-to-dark ratio of approximately 20:1, which is not considered sufficient. The scene at right has a low light-to-dark ratio of approximately 4:1, which maximizes the ability to assess an alarm.

A scene light-to-dark ratio of 6:1 should be considered to be a maximum (i.e., the lightest part of the scene being examined should be no more than 6 times brighter than the darkest part of the scene). A design ratio of less than 4:1 is strongly suggested for exterior lighting. Historically, assessment lighting systems have been designed to produce a 4:1 light-to-dark ratio at installation to allow degradation to a 6:1 ratio as bulbs produce a lower lumen output during their service life. At least 75 percent of the camera's field of view should have an even illumination that meets the minimum average illumination and light-to-dark ratio requirements.

Using the lens iris control, exposure time, and electronic signal amplification, a camera averages the total scene brightness detected by the imager. As a result, there is a limit to the amount and intensity of bright areas and dark areas in the camera's field of view for which the camera can compensate. This is known as the camera's **dynamic range**. Bright spots in the camera's field of view raise the average imager illumination level, causing the camera electronics to compensate by lowering the average video signal output. This tends to cause the darker portions of the image to become too dark and will negatively impact an operator's ability to assess the scene. To mitigate camera dynamic range limitations, a maximum scene light-to-dark ratio should be specified and implemented in the lighting design. If the light-to-dark ratio is excessive, light areas will provide too much light, causing a saturation of details within those areas. Likewise, dark areas will not provide sufficient light for good resolution.

Therefore, to maximize video assessment performance, the scene light-to-dark ratio should not exceed 6:1. This includes the entire area observed by the camera and not just the area of interest. To accomplish this, the lighting should extend beyond corners and fences to provide even illumination within 75 percent of the camera's field of view. As discussed previously, the outdoor perimeter area should use ground cover materials that provide a 25 to 35-percent reflectance.

6.4.6 Range of Scene Illumination Readings

If a minimum average scene illumination of 1.0 f-c and a 4:1 scene light-to-dark ratio is recommended at installation, a maximum and minimum set of illumination readings at the scene can be estimated. Table 6 shows a relative range of minimum to maximum illumination readings for average illumination values of 1.0 to 3.0 f-c. For example, given a normal distribution of illumination readings, if an average scene illumination is calculated to be 1.0 f-c, the individual illumination readings are estimated to be in the range of 0.5 to 2.0 f-c.

Table 6: Ranges for Average Illumination Values That Will Give 4:1 Light-to-Dark Ratios

Average Illumination (f-c)	Relative Range (f-c)
1.00	0.5–2.0
1.25	0.6–2.5
1.50	0.8–3.0
1.75	0.9–3.5
2.00	1.0–4.0
2.25	1.1–4.5
2.50	1.3–5.0
2.75	1.4–5.5
3.00	1.5–6.0

With an understanding of the relative levels of scene illumination produced by various sources and the amount of light reflected from typical scenes, the expected effectiveness of the camera portion of the video assessment system can then be determined. The next step is to understand the camera imaging system being used for alarm assessment.

6.4.7 Assessment Sensitivity

Lighting for video alarm assessment must consider the camera, video processing, and the display system that produces images for security personnel to view. Whether assessment is accomplished by security personnel or video cameras, lighting is required to facilitate 24-hour alarm assessment.

Since cameras are the primary mode of alarm assessment, camera sensitivity and lens aperture are two critical factors that affect the amount of reflected scene illumination appearing at the camera's imager. Camera sensitivity is defined as the amount of illumination required at the camera's imager to produce a usable video image. This basic rule applies whether the camera is analog or digital. The following factors contribute to producing a usable video assessment image:

- Illumination and the flatness of illumination present at the scene

- Spectral distribution (color of light) of the illumination source

- Total scene reflectance

- Object reflectance

- Camera lens transmittance

- Camera lens aperture diameter (f-stop)

- Spectral response of the camera imager (camera imager sensitivity to color of light)

- Camera imager sensitivity to broad spectrum light

- Camera automatic brightness control (gain and exposure time)

Under low light conditions, most cameras automatically compensate for the lack of illumination by increasing some combination of both the exposure time and amplifier gain depending on the overall brightness level desired by the user. Cameras with auto-iris lenses compensate for changing scene light levels by opening or closing the lens iris. The mechanics of illumination level compensation for some digital cameras can be programmed. The sequence of iris control, shutter control, and amplifier gain can be prioritized as to the order in which they are used to control imager illumination. Using shutter control for low light compensation causes the shutter to be open for longer exposure times. Long exposure times will blur moving objects while higher amplifier gain on very low light signals will produce grainier images. Both of these outcomes are undesirable for purposes of alarm video assessment. For example, fast-moving objects start to become noticeably blurry at exposure times greater than 1/6 of a second. The level of noise or amount of graininess in the monitor's video image that is acceptable depends on the application and the subjective opinion of the alarm station operator. Most digital cameras allow for limits to be placed on the maximum exposure time and gain so that one parameter can be favored over the other. Typically, some image blur can be tolerated, rather than picture graininess.

An industry standard for specifying camera sensitivity has not been established. Camera manufacturers often state camera sensitivity using varying test conditions and camera parameter settings. Often, camera sensitivity is stated in terms of minimum illumination level at the camera's imager to provide a usable picture. These camera specifications do not account for the illumination level degradation caused by the camera lens. While the amount of light needed at the camera's imager to produce a usable picture is specified, the amount of light that needs to enter the lens to achieve that light level at the imager may be significantly higher. Also, illumination and scene conditions during which the camera's sensitivity is determined may not necessarily be documented in the camera's specifications. Along with minimum light level, it is important to have the following information:

- Condition of output video—camera output and/or gain and exposure time

- Lens transmissivity—the percentage of incident light appearing at the front of the lens that passes through on to the imager

- Lens f-stop—the level of light reduction to the imager determined by the lens iris (or aperture) opening

- Test scene reflectance—the percentage of incident light on a scene that is reflected back to its source

In some cases, the parameters used to claim sensitivity may be unrealistically assumed to indicate a better performance than will be experienced in actual installations. For example, three of the favored parameters for specification enhancement are (1) higher scene reflectance than normally encountered, (2) unacceptably long exposure times, and (3) a large lens aperture (low f-stop). These parameters are usually determined with the camera viewing a static scene. If the need to effectively observe motion is factored into the specification process, the actual camera sensitivity experienced would most likely be less than (i.e., not as good as) that stated on manufacturers' data sheets.

To calculate whether the scene provides sufficient camera imager illumination for the camera chosen, the formula shown below may be used. To calculate the amount of illumination from the viewed scene incident on the camera imager, the scene illumination, scene reflectance, lens f-stop, and lens transmittance must be known. The formula below is a quick way to check if there will be sufficient illumination on the camera's imager surface to provide an adequate video picture. It is highly recommended, however, that a camera be field tested for verification under expected lighting conditions before its deployment.

$$\text{Imager Illumination} = \frac{\text{Scene Illumination x Scene Reflectance x Lens Transmittance}}{4 \text{ x (Lens f-stop)}^2}$$

where:
Imager illumination is in foot-candles or lux
Scene illumination is in foot-candles or lux
Lens f-stop from lens specification
Scene reflectance is in percent
Lens transmittance is in percent

For example, given an illuminated area and camera lens with the parameters shown below:

Scene Illumination = 1 f-c
Scene Reflectance = 0.3 (30%)
Lens Transmittance = 0.8 (80%)
Lens Focal Length = 1.8

Using the equation above:

$$\textit{Imager Illumination} = \frac{1 \times 0.3 \times 0.8}{4 \times (1.8)^2} = 0.019 \text{ f-c} \quad \text{or} \quad 0.19 \text{ lux}$$

From the example calculation, if a camera selected for video assessment had a sensitivity of 0.05 lux, it would have more than adequate sensitivity for the application if used to assess an area with the scene illumination and reflection specified and the lens f-stop and transmittance shown above. The scene provides 0.19 lux and the camera requires a minimum of 0.05 lux. Therefore, the scene provides 0.14 lux more than the minimum camera sensitivity.

Since most lighting applications are associated with bulbs that produce visible light, errors can be made in estimating the light required if the camera imager spectrum is not considered. An example of this relates to the use of color cameras. Most color cameras render color by masking pixels on an otherwise monochrome imager with red, green, and blue filters. In a typical camera imager configuration, 25 percent are red, 25 percent are blue, and 50 percent are green. The light spectrum is therefore filtered before it reaches the photo sensor resulting in degradation to the imager's overall pixel sensitivity when compared with monochrome cameras. Additionally, only 25 percent of the pixels in a color imager are sensitive to infrared light, as compared to 100 percent of the pixels in a black-and-white camera. A color camera's imager also has a blue filter in front of it that will further reduce its infrared sensitivity. If a near-infrared energy source is used, a black-and-white camera provides a visible video image even if the near-infrared illumination was not visible to the human eye.

6.4.8 Conceptual Lighting Layout

To prevent cameras from looking directly into a light source, security lights need to be located above the camera and out of the camera's field of view. It is usually recommended that security lights be no less than 3 vertical meters (10 vertical feet) above the camera's position. (See

Figure 75.) It is further recommended that a light source not be directly above the camera to mitigate the potential effects of dust- and fog-induced backscatter. However, the use of a sunshield that extends beyond the front cover glass of a camera enclosure like the brim of a cap can also minimize those effects.

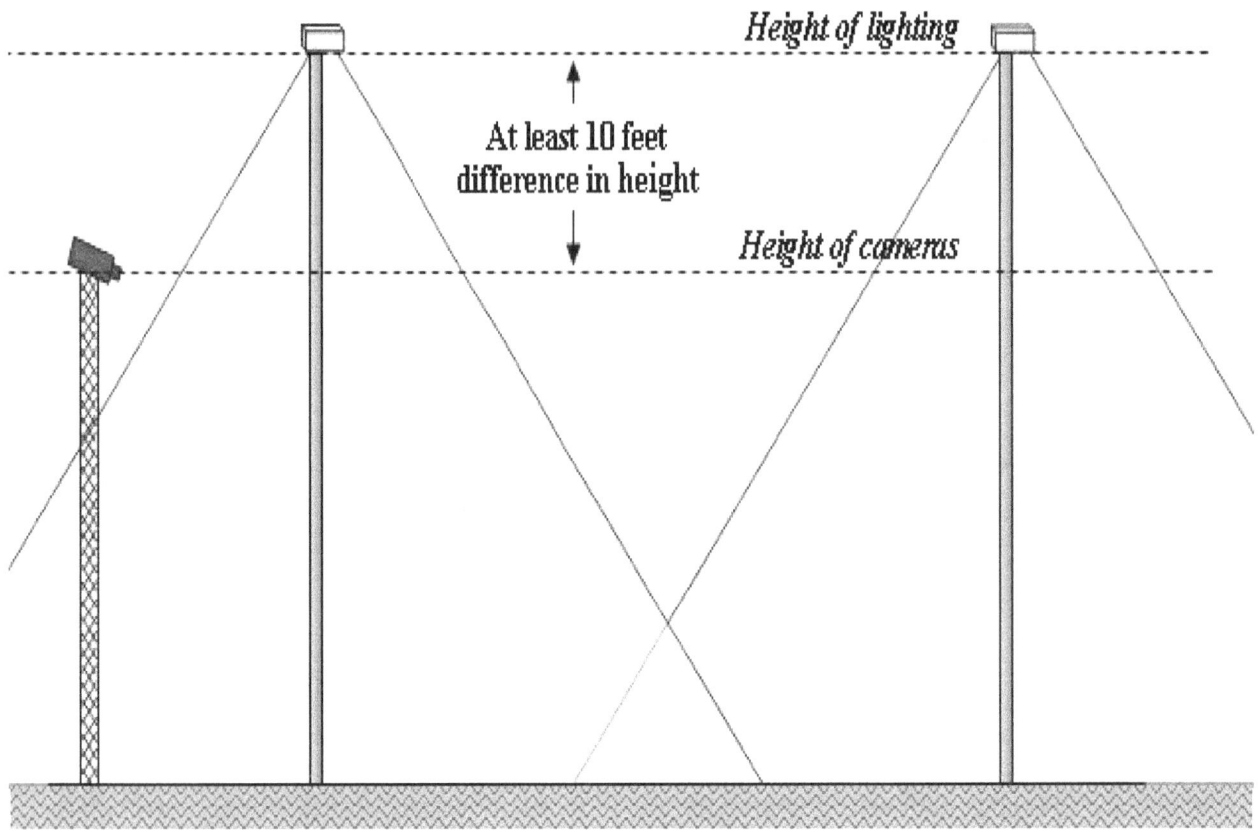

Figure 75: Recommended height differences between exterior lighting and cameras.

6.4.8.1 Lighting Design

Using the lighting guidelines discussed earlier, manufacturers and suppliers of chosen lighting equipment should be consulted to obtain luminaire and bulb illumination characteristics. Most major lighting suppliers and architectural engineering firms can model a proposed lighting system to determine and plot expected scene illumination levels. Maximum and minimum levels of illumination and light-to-dark ratio can also be determined. With access to such modeling software, several possible lighting configurations can be quickly evaluated. The inputs to the model normally include a file containing photometric data specific to a particular type of lamp (provided by the lamp manufacturer); luminaire light pattern (provided by the fixture manufacturer); and luminaire orientation, mounting height, and light pole spacing. A facility that is going to perform extensive lighting work should consider a lighting software program for in-house use; this type of software is generally not expensive. Figure 76 illustrates an example of output from an illumination modeling software package.

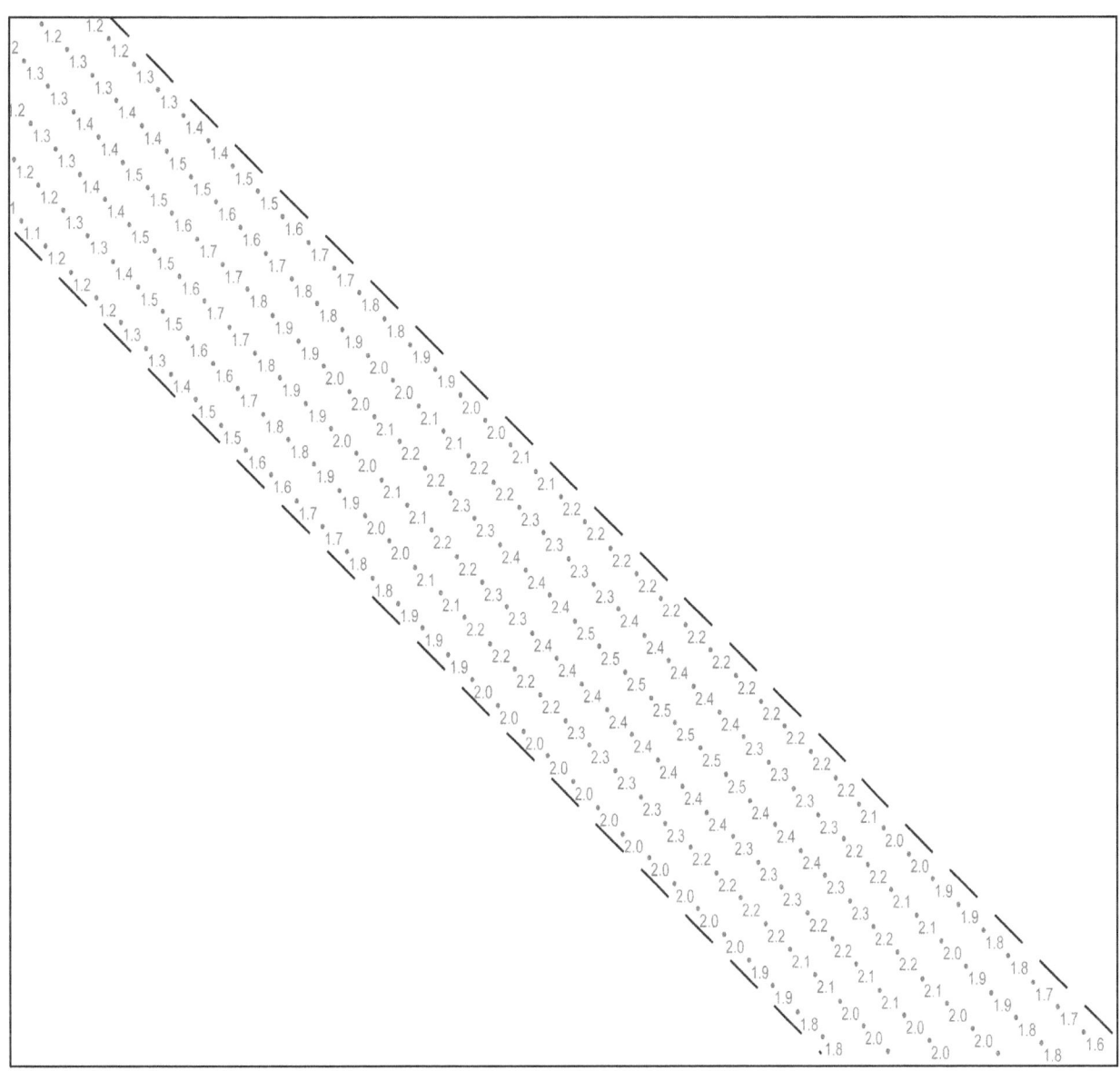

Figure 76: An example output from an illumination modeling software system.

The computer-generated modeling software illumination output pattern shown in Figure 77 provides data for a region of interest with expected illumination levels (i e , numbers within the dashed lines expressed in f-c) and other pertinent statistics. Note that the area calculated can represent data for a particular perimeter isolation zone.

The lighting design should be physically tested before installation in the final perimeter lighting configuration. This is accomplished by installing a minimum of five fixtures (for single-row installations) and actually measuring the illumination provided. Measurements should be made at 6 to 12 inches above ground level, taking measurements at 10-foot increments in the length and width dimensions. The brightest and darkest measurements should be identified, average illumination level calculated, and light-to-dark ratio calculated. Measurements should extend to distances where the light level is above 0.2 f-c.

Light measurements are taken using meters equipped with a photo sensor operating in the appropriate wavelength. Some light meters are configured so that the photo sensor can be interchanged to allow for measuring illumination within a specific spectral wavelength band. Measurements are typically taken 6 inches to 1 foot above the ground with the photo sensor facing up.

Adjustment and modification of the total lighting system after installation should be anticipated. It is not uncommon to discover during camera testing that there are unidentified reflections or bright spots that require correction.

The horizontal and vertical illumination levels required for interior lighting are determined similarly to exterior illumination values. The geometry of walls, floor, and ceiling reflections will affect ambient light levels.

Any interior lighting level suitable for human comfort and safety (e.g., between 30 and 100 f-c) will allow the use of a video camera with less sensitivity than is required for exterior video assessment applications. Energy savings can also be attained in interior spaces by illuminating a darkened room with near-infrared illumination sources and black-and-white cameras. The use of near-infrared illumination sources also provides a measure of covertness for the video assessment function in interior locations.

The wiring circuit for a lighting system should be arranged so that failure of any one lamp will not leave a large portion of the perimeter or a critical location in darkness. Electrical feeder lines should be placed underground to minimize the possibility of sabotage or vandalism from outside the perimeter. Another advantage to underground wiring is reduced effects from adverse weather conditions.

6.4.8.2   Lamp Types and Characteristics

Primary lamp types include the following:

- **Incandescent:** Light is emitted from a heated filament inside an evacuated globe.

- **Quartz Iodine:** Light is generated as in an incandescent lamp, but the globe is filled with a halogen gas that regenerates the filament and allows higher intensity.

- **Fluorescent Lamp:** Light is generated by an electric arc in a tube filled with mercury vapor. The low-pressure vapor emits ultraviolet radiation that is converted to visible light by fluorescent powders on the inner surface of the tube.

- **High-Intensity Discharge Lamp:** The light energy is generated by direct interaction of an arc with the gas to produce visible light. High-intensity discharge lamps include mercury vapor lamps, metal halide lamps, and high- and low-pressure sodium lamps. Argon is normally added to aid starting, and various powders or vapors may be added to improve color rendition.

- **Light Emitting Diode (LED):** The light energy is generated based on the light emitting diode solid-state technology. Lamps are usually constructed in cluster LEDs within suitable housing. Today, the LED lamp is mostly used for infrared illumination. However, LED may become more of a standard in the future for all types of lighting when considering the following advantages it offers over conventional lighting:

- High efficiency—LEDs are available that reliably offer over 100 lumens per watt.
- Life span—If properly engineered, LEDs can operate 50,000 to 60,000 hours.
- Instant restrike—Lamps illuminate upon activation without delay.
- Mercury free—Unlike fluorescent and most HID technologies, LEDs contain no hazardous mercury or halogen gases.
- Tunable spectrum—LEDs can be tuned to emit light in a broad range of colors.

Table 7 summarizes lamp characteristics for seven common bulb types.

Table 7: Characteristics of Seven Common Types of Bulbs

	Incandescent	Quartz Iodine	Mercury Vapor (fluorescent)	Metal Halide	High-Pressure Sodium	Low-Pressure Sodium	LED
Lamp Efficiency (lumens/ watt)	12–20	20–23	40–65	80–100	95–130	131–183	70–100
Approx. Life Span (hours)	750–10,000	2,000	24,000	15,000	20,000	18,000	50,000 – 60,000
Time Required for Full Output	Immediate	Immediate	3–7 minutes	3–5 minutes	3–4 minutes	8–15 minutes	Immediate
Spectrum	Broad (Visible to NIR ~ 400 to 1,000 nm)	Broad (Visible to NIR ~ 400 to 1,000 nm)	Blue Green	Broad (Visible to NIR ~ 400 to 1,000 nm)	Gold-Yellow	Mono-chromatic Yellow	Broad (Visible to NIR ~ 400 to 1,000 nm)

6.4.9 Maintenance for Lighting Systems

Periodic inspections should be made of all electrical circuits to replace or repair worn parts, tighten connections, and check insulation. Fixtures should be kept clean and correctly aimed to provide optimal service.

Primary and alternate power sources should be identified. The following is a partial list of considerations:

- The primary source is usually utility offsite power.

- An alternate source such as supply UPS consisting of batteries and diesel-fuel-driven generators is provided where required and should do the following:

  - Provide required power automatically, without interrupting assigned illumination upon failure of primary power
  - Be adequate to power the entire lighting system

- Be equipped with adequate fuel storage and supply
- Be tested under load to ensure efficiency and effectiveness
- Be located within a controlled area or hardened building inside the perimeter for additional security

6.4.9.1   Measuring Perimeter Illumination Levels

Facilities should adopt a standard method for taking initial and periodic perimeter illumination level measurements. This standard measurement will ensure that the light levels can be correlated to previous measurements for the purpose of determining the amount of illumination degradation as bulbs age. Tracking the illumination degradation of bulbs will allow maintenance personnel to predict when luminaires need to be cleaned and lenses or bulbs replaced.

The measurement process requires two persons and the use of a 300-foot tape measure, a calibrated light meter, a 1-foot-high pedestal for attaching the light meter sensor, at least eight wooden blocks (see Figure 77), a clipboard, and a preprinted spreadsheet for recording measurements. Measurements are taken after sunset under darkened sky conditions after lights have been on sufficiently long to achieve full brightness. While the following instructions depict measurements for a 300-foot long area, the process can be applied to any length.

Figure 77: Checking light levels using a light meter within a perimeter isolation zone during nighttime illumination.

6.4.9.2   Preparation for Measuring Perimeter Illumination Levels

The following procedure should be used to prepare for measuring perimeter illumination levels:

(1)   Beginning directly underneath one luminaire at the inner fence location (which will be called 0′ width and 0′ distance) and using a tape measure, place wooden blocks at 10-foot intervals across the width of the perimeter toward the outer fence. (See Figure 78.)

(2)   As in Step 1, but at a 300-foot distance from the initial starting point along the inner fence, using a tape measure, place wooden blocks at 10-foot intervals across the width of the perimeter towards the outer fence.

(3)   With two persons, one at each end of the tape measure, extend the tape measure the entire 300-foot length to the second wooden block along the inner fence. Ensure that the tape measure is "face up" for its entire length.

(4)   Firmly attach the light meter sensor (face up) to the top of the pedestal with duct tape.

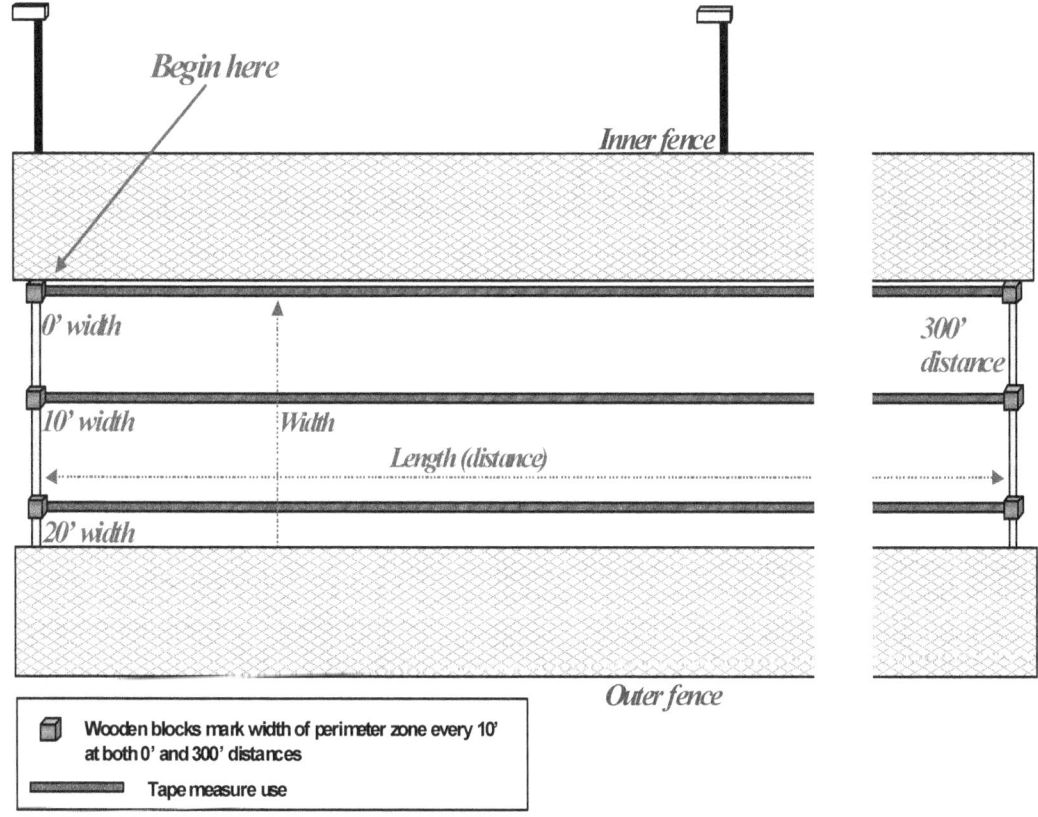

Figure 78: Preparation for measuring perimeter lighting illumination.

6.4.9.3 Procedure for Measuring Perimeter Illumination Levels

This procedure requires two persons; one person takes the measurement and the other person records the reading on a data sheet.

(1) Beginning at the first measurement point, place the pedestal and sensor at the zero measurement point on the tape measure along the inner fence under the light fixture.

(2) Read the illumination value from the light meter. Ensure that personnel or other obstructions are not casting a shadow on the light meter sensor.

(3) Record the light meter reading on a data sheet (see a sample data sheet in Table 88).

(4) Move to the 10-foot mark on the tape measure, and read and record the light meter reading on the data sheet.

(5) Move to the 20-foot mark on the tape measure, and repeat the measurement and recording procedure.

(6) Continue the measurement procedure until the 300-foot mark on the tape measure is reached.

(7) Two persons, one on each end of the tape measure, move the tape measure to the two wooden blocks at the 10-foot distance from the inner fence.

(8) Take measurements from the zero mark to the 300-foot mark as described in Steps 2 through 7.

(9) Move the tape measure to the blocks at the 20-foot and 30-foot distances (depending on the zone's width) from the inner fence, and repeat the procedure described in Steps 2 through 7.

(10) When all the readings are taken, sum them and divide by the number of readings. For the example shown in Figure 79, the number of readings taken is 124 (31 readings per row multiplied by 4 rows). This is the *average illumination level*.

(11) Ensure that the average illumination is greater than 1.0 f-c or 10 lux.

(12) Find the highest reading and lowest reading on the data sheet. Divide the highest reading by the lowest reading; this is the *light-to-dark ratio*.

(13) Ensure that the light-to-dark ratio is less than 6:1 (should be approximately 4:1 for initial installation readings).

(14) As a second check of readings, divide the average illumination level, calculated in Step 11 above, by the lowest reading; this is the *average-to-dark ratio*. A well-designed lighting system will have an average light-to-dark ratio of approximately 2.5-to-1 (2.5:1) or less.

(15) Document the readings and calculations for comparison to subsequent readings.

(16) Compare calculations taken at initial installation with subsequent readings to determine illumination degradation.

(17) Plot initial and subsequent average illumination, light-to-dark ratio, and average-to-dark ratio values after each set of measurements to show long-term trends in lighting degradation.

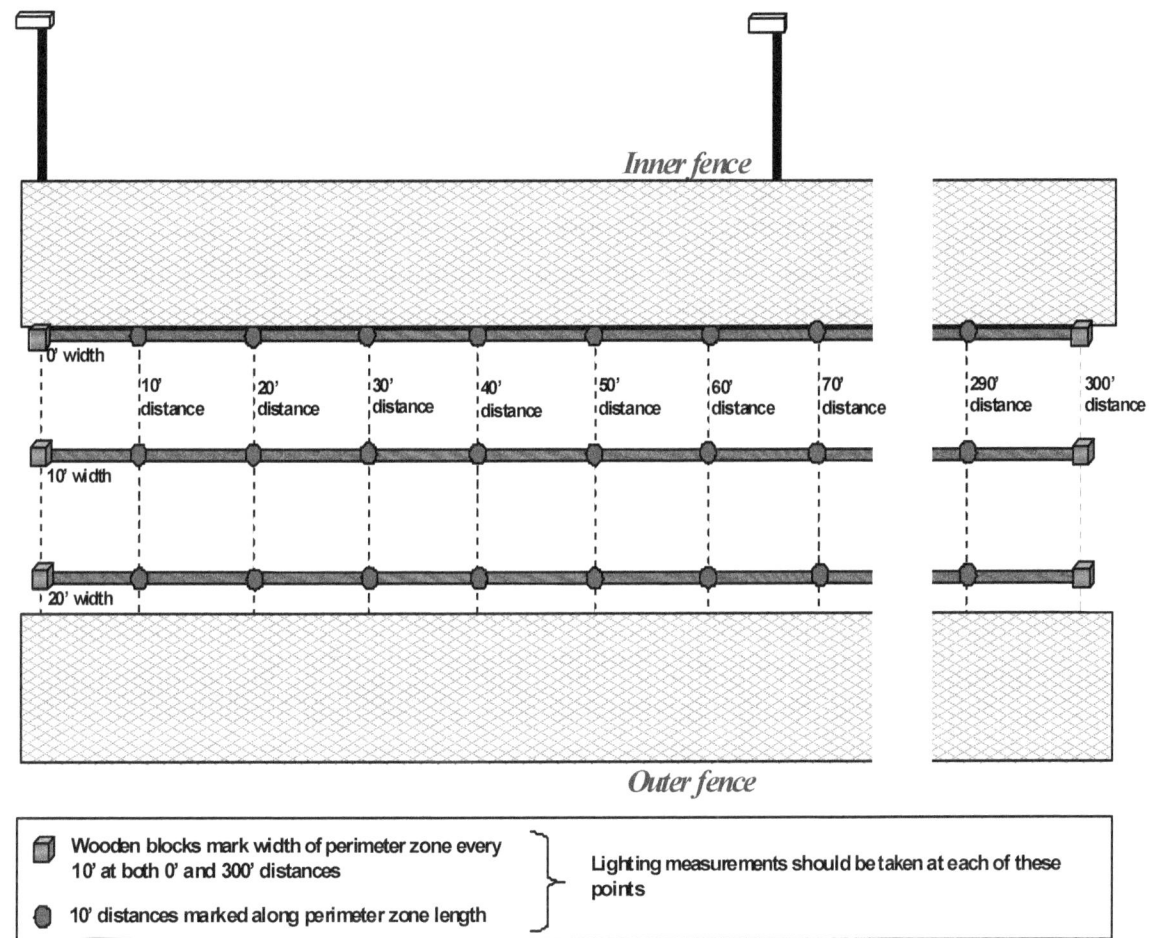

Figure 79: Points at which to measure light readings.

Note: Every wooden block or red circle from the preceding diagram should have a coordinating measurement within one of the squares in the table on the next page.

Table 8: Perimeter Light Measurements

	Width of Perimeter (as appropriate)					
	0′	10′	20′	30′	40′	⊗
0′						
10′						
20′						
30′						
40′						
50′						
60′						
70′						
80′						
90′						
100′						
120′						
130′						
140′						
150′						

LENGTH OF PERIMETER

Mark locations of lights

6.5 Video Assessment System Testing

Testing requires that all the elements of the video assessment system work as integrated components, not individual parts. Therefore, each aspect of the video assessment system should be tested in concert. Video assessment systems that are integrated with intrusion detection systems should be tested in conjunction with each other as it is important to verify that these integrated systems are functioning and performing in the manner that they are interfaced.

A regular program of testing system components is imperative for maintaining them in optimal operating order. Three types of testing need to be performed at different times in the life of a video assessment system: acceptance testing, performance testing, and operability testing.

6.5.1 Acceptance Testing for Video Assessment Systems

Upon receiving cameras for final installation, camera performance should be evaluated to determine conformity with the manufacturer's specifications, compatibility with design criteria, and consistent performance from camera to camera. Experience has shown that final inspection at the manufacturer's plant may produce cameras with nonequivalent performance for specific field applications and lighting conditions. On occasion, equipment arrives in damaged condition, or parts have shaken loose in transit. Operating equipment continuously (performing burn-in) for a few hundred hours before final installation will decrease issues associated with immediate equipment mortality after initial installation.

Exterior cameras should be installed in accordance with the manufacturer's requirements and focused at night while exposed to nighttime lighting conditions. Some cameras are shipped prefocused to specified distances. However, the environment in which these cameras are focused may not be the same as the operating environment at the facility. Initial camera setup should be followed by indoor and outdoor testing to confirm that the cameras perform as expected. Final adjustments to camera focus and video image adequacy under nighttime lighting conditions should be made. The video images produced by the cameras should not vary much in picture quality between daytime and nighttime operation. Washed-out images during the day or dark, low-contrast images at night are indications that cameras are not performing to intended requirements.

Cameras should be checked for resolution and complete coverage of the assessed zone to ensure that there are no detection areas in a zone that the camera cannot see. One simple method of checking camera resolution is to use 1-foot-sized triangular, circular, and square targets. The targets, painted white on one side and black on another, are placed at the end of an assessment zone, and an alarm station operator should correctly identify the target shapes. One-foot targets simulate the frontal cross-section of a crawling person. The use of a large field resolution chart such as a Rotakin can also be used to provide more quantifiable information of camera resolution. In addition, the alarm station operator should be able to see the feet of an individual standing at both the inner and outer fence when the operator is standing at the beginning of the assessed zone. This test should be performed on live video from the cameras, as well as on playback video from the video recording device for each assessment zone. In interior spaces, particular attention should be paid to the location of equipment or other objects that might occlude camera view or create shadows or blind spots.

Video system speed of response should be tested to ensure that alarm assessment and video recording of an intrusion detection zone occurs within 1 or 2 seconds of alarm annunciation.

Other performance tests include a determination of the maximum number of concurrent alarms that can be processed, logged, and recorded at the same time and the amount of assessment delay occurring if more than one alarm occurs concurrently. The desired specifications and statement of acceptance testing requirements should be included as part of the terms and conditions of system purchase from the security systems integrator.

Conducting equipment incoming inspection, burn-in, and adjustments before actual use should minimize maintenance and failure problems in the short term. A dated maintenance log should be kept to document occurrences of problems, problem resolution, and if recurring problems occur, long-term fixes or equipment replacements undertaken to remediate repetitive problems. Maintenance cycles can be established for the performance of repetitive maintenance activities and equipment replacement activities as a result of the data collected on the performance and failures occurring during system operation. This practice will substantially reduce repair time and identify substandard equipment performance.

Equipment spares should be planned for at the design phase of the video assessment system so that failures can be immediately replaced. The spares inventory should be replenished as spare parts are put into service.

One copy of the manufacturers' equipment documentation should be kept in a central document storage location and another copy should be kept by the security equipment maintenance organization. Modifications to the security system (as initially installed or documented) should be documented and stored at these two locations.

Acceptance testing for the video assessment system is the most encompassing because baseline performance and operability are determined and documented. Acceptance tests will uncover operational and functionality issues that need to be addressed to ensure system operation in accordance with design specifications.

Use the following guidelines to perform acceptance testing:

6.5.1.1   Cameras and Digital Video Recorders Used To Play Back Camera Images

(1)   Ensure that each camera produces a video image on the alarm station monitors and that each camera channel produces video playback images for that channel on the alarm station monitors.

(2)   Ensure that camera channel numbering is arranged in a logical rather than haphazard order. For example, cameras for adjacent perimeter sectors should be sequentially numbered, and cameras inside buildings should be numbered according to a logical flow traversing through the building. For DVRs, ensure that the numbering of DVR camera input channels agrees with the live camera channels indicated on the alarm station monitors.

(3)   With multiple individuals in the field communicating with alarm station operators, ensure that the camera image displayed on the monitor and recorded on the digital recorder is the correct one for the assessed space indicated. The alarm station operators should observe the person in the field. Also, ensure that playback of video from the recorder shows the correct field location for the camera channel being tested.

(4)   If camera images have graphic legends displayed on the monitor, ensure that the graphic legend is labeled correctly for the camera channel being tested.

(5)   Ensure that camera resolution is sufficient for intruder classification and that cameras are in focus at night with nighttime illumination. To perform this test, one individual in the field places a 1-foot triangle, circle, and square at ground level at the end of the assessment zone or area. The alarm station operator, communicating with the individual in the field via two-way radio, should correctly identify the order of the triangle, circle, and square. Alternately, using a field resolution chart calibrated for the distance from the camera, the alarm station operator should be able to accurately observe four distinct lines on the resolution chart at the calibrated position on the chart.

(6)   Ensure that video recorder playback resolution is sufficient for intruder classification. Repeat the procedure for determining camera resolution (in the preceding step) to ensure video recorder playback resolution.

(7)   Ensure that exterior cameras are focused so that the camera image is "in focus" at both the near and far fields of view. For example, check to ensure that the video image at both the beginning and end of an assessment zone are in focus.

(8)   In exterior perimeter isolation zones, ensure that cameras are aimed so that the entire perimeter width can be observed on the alarm station monitors and in the recorded video. Camera aim can be tested by using two orange cones and two fiberglass rods in the following fashion. Two 2-foot-long fiberglass rods with a reflector at one end are fabricated with a clamping mechanism (like a battery jumper cable clamp) for attaching to the top horizontal pipe of the perimeter fence (rods are available at a local home building materials store). The rods are attached to the top of the inner and outer perimeter fences at the beginning of each perimeter isolation zone. The two cones are placed along the perimeter side of the inner and outer perimeter fences at the beginning of each assessment zone. The alarm station operator should be able to see the bottoms of the cones at the beginning of the sector, as well as the reflectors at the top of the fence attachment rods.

(9)   Ensure that a low-profile human observed on the alarm station monitors (and on the playback of recorded video) can be classified at the far end of each perimeter assessment zone. This should be tested at night, at the inner and outer fence line inside the perimeter and at the center of the perimeter. The test subject should be a small individual who performs a belly crawl with his or her head toward the camera. The alarm station operator observing the monitor (and the playback of recorded video) should be able to determine that the simulated intruder can be observed and accurately classified.

(10)   During bright sunlight and during nighttime illuminated conditions, observe each exterior camera on the alarm station monitors (and on the playback of recorded video), and ensure that images from each camera have approximately the same brightness and contrast and that monitor images of assessment zones do not have bright spots or dark spots. For images that are too bright, too dark, or lacking in contrast, camera or lens adjustment or replacement may be required. For analog camera systems, video transmission or communication modules may also be the source of video brightness anomalies.

(11)   Observe the scene from each interior camera on the alarm station monitors and ensure that images from each camera have approximately the same brightness and contrast.

For images that are either too bright, too dark, or lacking in contrast, camera or lens adjustment or replacement may be required. For analog camera systems, video transmission or communication modules may also be the source of video brightness anomalies.

(12) Observe interior and exterior cameras on the alarm station monitors and ensure that images are clear and crisp and do not have fuzzy or flickering images.

6.5.1.2 Camera Tests

(1) Ensure that exterior cameras are tilted down and do not view above the horizon at the camera's far field of view.

(2) Ensure that bright spots, shiny reflections, or glare from luminaires does not appear in or cast a bright image in a camera's field of view.

(3) Ensure that large objects, such as electrical junction boxes that a person could hide behind, are not in an exterior camera's fields of view.

(4) Check all fasteners for the camera tower and mount to ensure they are secure and there is no movement caused by loose fasteners.

(5) Ensure that the camera, enclosure, and mount are firmly affixed and not affected by wind causing movement of camera.

(6) For camera enclosures with external sunshades, ensure the enclosures have sunshades that extend at least 2 inches beyond the front of the camera enclosure.

(7) During blowing snow conditions, ensure that enclosure heaters melt the snow accumulation on the enclosure front cover glass without creating an accumulation of ice on the front cover glass.

6.5.1.3 Intrusion Alarm and Assessment Camera Interface Tests

Ensure that intrusion alarms in assessed areas trigger the appropriate camera's video to appear on the alarm station monitors (and on the playback of recorded video). While communicating with the alarm station operator using a two-way radio, an individual in the field should trigger an alarm in each intrusion detection zone. The alarm station operator should ensure that the assessment video displayed because of the alarm is from the correct camera for the zone in alarm and that the assessment video appears on the monitor within seconds. If an intrusion detection zone has multiple sensors, an intrusion alarm and video verification should be performed for each sensor.

6.5.1.4 Camera Lighting Tests

Turn off nighttime illumination, and ensure that a "loss of video contrast" alarm occurs for each camera.

6.5.1.5   Power Tests

(1)   Disconnect or switch the main source of power to the video assessment system to simulate loss of offsite power, and ensure that the UPS and diesel generator maintain uninterrupted power to the intrusion detection and assessment systems for the appropriate timeframe (will vary with the facility.)

(2)   Disconnect individual cameras at field camera junction boxes, and ensure that a "loss of video" alarm occurs for each camera.

6.5.1.6   Tampering and Accidental Disconnection Tests

(1)   Cover the front of the camera enclosure, and ensure that a "loss of video contrast" alarm occurs for each camera.

(2)   Shine a bright light into the front of each camera, and ensure that a "loss of video contrast" alarm occurs for each camera.

(3)   Disconnect the video signal cable from each analog camera, and ensure that a "loss of video" alarm occurs for each camera.

(4)   Disconnect the fiber-optic video transmission cable (fiber) for each camera, and ensure that a "loss of video" alarm occurs for each camera.

(5)   Disconnect the Ethernet cable from each digital camera, and ensure that a "loss of video" alarm occurs for each camera.

(6)   Disconnect the camera Ethernet cable to Ethernet switches carrying video signals, and ensure that a "loss of video" alarm occurs for all cameras connected to the switch.

(7)   Disconnect the power cable to Ethernet switches carrying video signals, and ensure that a "loss of video" alarm occurs for all cameras connected to the switch.

(8)   Simultaneously create intrusion alarms in multiple (two through five) adjacent and nonadjacent perimeter intrusion detection zones. Ensure that the alarm assessment video cues the recorded and live (pre- and post-alarm) video and displays the video for each of the alarming zones.

6.5.1.7   Camera Towers

Ensure that camera towers are properly grounded and have a lightning air terminal at the top of the tower. Also ensure that the tower and lightning rod (air terminal) have cables exothermically bonded to a ground rod at each tower. Resistance from the lightning rod and the tower to the ground rod cable attaching point must be less than 1 ohm.

6.5.1.8   Camera Tower Junction Boxes and Field Distribution Junction Boxes

Ensure that the camera tower junction box and field distribution junction box create tamper alarms when each enclosure is opened. Also ensure that the alarm station tamper alarm notification indicates the opening of the correct junction box. To accomplish this test, an individual in the field communicates with the alarm station operator using a two-way radio. The individual in the field opens each camera tower junction box and field distribution panel (one at a time). The alarm station operator verifies that the correct tamper alarm message occurs for the box opened. After each test, the junction box door is reclosed. The alarm station operator observes that the tamper alarm can be cleared after the junction box door is closed.

6.5.1.9   Shadow and Light Variation Determinations

(1)   During nighttime illumination of perimeter isolation zones, observe monitor images of each assessed zone to ensure that fences are vertically perpendicular to the ground so that a shadow alongside the base of the perimeter side of the fence is not present.

(2)   Ensure that buildings and large equipment enclosures do not cast a shadow across the perimeter isolation zone during either daytime or nighttime hours. Shadow areas reduce image contrast and video assessment capability.

6.5.1.10 Camera Brightness

(1)   Ensure that nighttime bright light illumination from adjacent buildings or parking lots does not create bright spots on the floor of the perimeter.

(2)   Ensure that camera brightness is consistent during changing daylight brightness conditions, particularly during morning and evening hours. This ensures that camera electronics and auto-iris lens controllers are compensating properly for changing illumination levels. This condition is normally observed as a video image that is either brighter or darker than other camera images. For images that are too bright, too dark, or lacking contrast, camera or lens adjustment or replacement may be required. For analog camera systems, video transmission or communication modules may also be the source of video brightness anomalies.

(3)   Ensure that camera brightness is not oscillating between two auto-iris positions. This is observed as an oscillating lighter to darker camera image. This effect can be caused by dust in the auto-iris lens mechanism or improper combination of lens control and camera amplifier electronics parameter adjustments. Camera or lens adjustment or replacement may be required.

6.5.2 Operability Testing

Operability testing for the video assessment system is an ongoing set of tests to ensure that the video assessment system continues to function properly. Operability tests identify sources of system degradation and nonfunctionality requiring immediate attention and remediation. Tests are to be conducted for all camera locations. It is recommended that these tests be performed in conjunction with intrusion detection system and component tests to verify that both systems are operating and interfacing as required. Several of these tests repeat the acceptability tests.

(1)   Ensure that intrusion alarms in assessed areas cause the appropriate camera's video to appear on the alarm station monitor (and on the playback of recorded video). While communicating with an alarm station operator using a two-way radio, an individual in the field triggers an alarm in an intrusion detection zone. The alarm station operator ensures that the assessment video resulting from the alarm is from the correct camera for the zone in alarm and that the assessment video appears on the alarm station

monitors within seconds. If multiple sensors are in an intrusion detection zone, an intrusion alarm is created, and video verification is performed for each sensor.

(2)   Ensure that the graphic legend is labeled correctly for the alarming camera channel being tested, and ensure that the graphic legend is stable and not "jittering" on the monitor screen.

(3)   Ensure that interior cameras produce an image that is "in focus." Ensure that exterior cameras are "in focus" at both the near and far fields of view. Ensure that the video image at both the beginning and end of an assessment zone are in focus. Ensure that camera images are clear and crisp and do not have fuzzy or flickering images.

(4)   Ensure that camera focus and video recorder playback quality is sufficient for human intruder classification. One individual in the field places a 1-foot triangle, circle, and square at ground level at the end of the assessment zone or area. The alarm station operator should correctly identify the order of the triangle, circle, and square by observing live camera video on the alarm station monitor and playback video images.

(5)   The following is an alternate method. At night, at the inner and outer fence line inside the perimeter and at the center of the perimeter, a small individual "on all fours," with his or her head towards the camera, moves sideways in the camera field of view. An alarm station operator observing the monitor in the alarm station should be able to determine that the simulated intruder can be observed and accurately classified.

(6)   In exterior perimeter isolation zones, ensure that cameras are aimed so that the entire perimeter width can be observed on the alarm station monitor. For this test, two orange cones are required. The two cones are placed along the perimeter side of the inner and outer perimeter fences at the beginning of each assessment zone. The alarm station operator should be able to see the bottoms of the cones at the beginning of the sector.

(7)   For interior cameras and for exterior cameras, during bright sunlight and during nighttime illumination conditions, observe camera images on the alarm station monitor and ensure that camera images have approximately the same brightness and contrast and that monitor images of assessment zones do not have bright spots or dark spots. For images that are too bright, too dark, or lacking in contrast, camera or lens adjustment or replacement may be required. For analog camera systems, video transmission or communication modules may also be the source of video brightness anomalies.

(8)   Ensure that camera brightness is not oscillating between two auto-iris positions. This is observed as an oscillating lighter-to-darker camera image. This effect can be caused by dust in the auto-iris lens mechanism or improper combination of lens control and camera amplifier electronics parameter adjustments. Camera or lens adjustment or replacement may be required.

6.5.3 Performance Testing

Performance testing for the video assessment system is a set of tests performed to ensure that the entire video assessment system is performing to design requirements and all functions are operating properly. Performance tests are an indepth set of tests to identify sources of system degradation and nonfunctionality needing repairs to bring the system back to initial performance specifications. Tests are to be conducted for all camera locations. It is recommended that these tests be performed on the entire set of camera-assessed areas on an annual schedule.

(1) Ensure that each camera produces a video image on the monitor (and on the playback of recorded video).

(2) Ensure that intrusion alarms in assessed areas cause the appropriate camera's video to appear on the alarm station monitor and to be recorded on the video recorder. While communicating with an alarm station operator using a two-way radio, an individual in the field triggers an alarm in an intrusion detection zone. The alarm station operator ensures that the live video and playback of recorded assessment video resulting from the alarm is from the correct camera for the zone in alarm and that the assessment video appears on the alarm station monitor within seconds. If multiple sensors are in an intrusion detection zone, an intrusion alarm is created, and video verification is performed for each sensor.

(3) Ensure that the graphic legend is labeled correctly for the alarming camera channel being tested, and ensure that the graphic legend is stable and not "jittering" on the monitor screen.

(4) Ensure that interior cameras produce an image that is in focus. Ensure that exterior cameras are in focus at both the near and far fields of view. Ensure that the video image at both the beginning and end of an assessment zone is in focus. Ensure that camera images are clear and crisp and do not have fuzzy or flickering images.

(5) Ensure that camera focus and video recorder playback quality are sufficient for human intruder classification. One individual in the field places a 1-foot triangle, circle, and square at ground level at the end of the assessment zone or area. While communicating with the individual in the field via two-way radio, an alarm station operator observes live camera video on an alarm station monitor and video recorder playback images and should correctly identify the order of the triangle, circle, and square.

(6) In exterior perimeter isolation zones, ensure that cameras are aimed so that the entire perimeter width can be observed on the alarm station monitor. For this test, two orange cones and two fiberglass rods are required. Two 2-foot-long fiberglass rods with a reflector at one end are fabricated with a clamping mechanism (like a battery jumper cable clamp) for attaching to the top horizontal pipe of the perimeter fence (rods are available at local home building materials stores). The rods are attached to the top of the inner and outer perimeter fences at the beginning of each perimeter isolation zone. The two cones are placed along the perimeter side of the inner and outer perimeter fences at the beginning of each assessment zone. The alarm station operator should be able to see the bottoms of the cones at the beginning of the sector, as well as the reflectors at the top of the fence attachment rods.

(7) Ensure that a low-profile human observed on an alarm station monitor can be classified at the far end of each perimeter assessment zone. At night, at the inner and outer fence line inside the perimeter, and at the center of the perimeter, a small individual performs a belly crawl with his or her head toward the camera. The alarm station operator observing the monitor should be able to determine that the simulated intruder can be observed and accurately classified.

(8) For interior cameras and for exterior cameras, during bright sunlight and during nighttime illuminated conditions, observe camera images on the alarm station monitor and ensure that camera images have approximately the same brightness and contrast and that monitor images of assessment zones do not have bright spots or dark spots. For images that are too bright, too dark, or lacking contrast, camera or lens adjustment or replacement may be required. For analog camera systems, video transmission or communication modules may also be the source of video brightness anomalies.

(9) Ensure that camera brightness is not oscillating between two auto-iris positions. This is observed as an oscillating lighter-to-darker camera image. This effect can be caused by dust in the auto-iris lens mechanism or improper combination of lens control and camera amplifier electronics parameter adjustments. Camera or lens adjustment or replacement may be required.

(10) Ensure that a low-profile human observed on the playback of video taken during Test 8 above can be classified at the far end of each perimeter assessment zone. At night, at the inner and outer fence line inside the perimeter, and at the center of the perimeter, a small individual performs a belly crawl with his or her head toward the camera. The alarm station operator observing the playback of belly crawl activities should be able to determine that the simulated intruder can be observed and accurately classified.

(11) During bright sunlight and during nighttime conditions, observe each exterior camera on the alarm station monitor and ensure that images from each camera have approximately the same brightness and contrast and that monitor images of assessment zones do not have bright spots or dark spots. For images that are too bright, too dark, or lacking contrast, camera or lens adjustment or replacement may be required. For analog camera systems, video transmission or communication modules may also be the source of video brightness anomalies.

(12) Observe interior and exterior cameras on an alarm station monitor, and ensure that images are clear and crisp and do not have fuzzy or flickering images.

(13) Ensure that exterior cameras are tilted down and do not view above the horizon at the camera's far field of view.

(14) Ensure that bright spots, shiny reflections, or glare from luminaires does not appear in or cast a bright image in a camera's field of view.

(15) Ensure that large objects, such as electrical junction boxes that a person could hide behind, are not in an exterior camera's field of view.

(16) Check all fasteners for the camera tower and mount to ensure they are secure and there is no movement caused by loose fasteners.

(17) Ensure that the camera, enclosure and mount are firmly affixed and not affected by wind causing movement of camera.

(18) During blowing snow conditions, ensure that enclosure heaters melt any snow accumulation on the enclosure's front cover glass without creating an accumulation of ice on the front cover glass.

(19) Turn off nighttime illumination, and ensure that a "loss of video contrast" alarm occurs for each camera.

(20) Disconnect or switch the main source of power to the video assessment system to simulate a loss of offsite power, and ensure that the UPS and diesel generator maintain uninterrupted power to the intrusion detection and assessment systems for the appropriate timeframe (will vary with the facility).

(21) Disconnect power to individual cameras at field camera junction boxes, and ensure that a "loss of video" alarm occurs for each camera.

(22) Cover the front of the camera enclosure, and ensure that a "loss of video contrast" alarm occurs for each camera.

(23) Shine a bright light into the front of each camera, and ensure that a "loss of video contrast" alarm occurs for each camera.

(24) Disconnect the video signal cable from each analog camera, and ensure that a "loss of video" alarm occurs for each camera.

(25) Disconnect the fiber-optic video transmission cable (fiber) for each camera, and ensure that a "loss of video" alarm occurs for each camera.

(26) Disconnect the Ethernet cable from each digital camera, and ensure that a "loss of video" alarm occurs for each camera.

(27) Disconnect the camera Ethernet cable to Ethernet switches carrying video signals, and ensure that a "loss of video" alarm occurs for all cameras connected to the switch.

(28) Disconnect the power cable to Ethernet switches carrying video signals, and ensure that a "loss of video" alarm occurs for all cameras connected to the switch.

(29) Simultaneously create intrusion alarms in multiple (two through five) adjacent and nonadjacent perimeter intrusion detection zones. Ensure that the alarm assessment video queues the recorded and live (pre- and post-alarm) video and displays the video for each of the alarming zones.

(30) Ensure that camera towers are properly grounded, have lightning air terminals at the top of each tower, and the tower and lightning rod (air terminal) have cables exothermically bonded to a ground rod at each tower. Resistance from the lightning rod and the tower to the ground rod cable attaching point must be less than 1 ohm.

(31) Ensure that the camera tower junction box and the field distribution junction box create tamper alarms when each enclosure is opened. Ensure that the alarm station receives the tamper alarm notification and indicates the opening of the correct junction box. An individual in the field communicates with an alarm station operator using a two-way radio. The individual in the field opens each camera tower junction box and field distribution panel (one at a time). The alarm station operator verifies that the correct tamper alarm message occurs for the box opened. After each test, the junction box door is reclosed. The alarm station operator observes that the tamper alarm can be cleared after the junction box door is closed.

(32) During nighttime illumination of perimeter isolation zones, observe monitor images of each assessed zone to ensure that fences are vertically perpendicular to the ground so that a shadow along the base of the perimeter side of the fence is not present.

(33) Ensure that buildings and large equipment enclosures do not cast a shadow across the perimeter during daytime or nighttime hours. Shadow areas reduce image contrast and video assessment capability.

(34) Ensure that nighttime bright light illumination from adjacent buildings or parking lots does not create bright spots on the floor of the perimeter.

6.6 Maintenance

A video assessment system requires regular maintenance to ensure a high level of performance and reliability.

Maintenance should be performed according to a defined schedule. If performance or operability testing show degraded system performance, the maintenance schedule should be adjusted accordingly.

Maintenance procedures and equipment needed to maintain the video assessment system should be considered during the design phase of the system. The long-term operability and performance of the system will depend on effective maintenance. Spare part purchases should be made at the same time as the primary equipment purchases to ensure video assessment system operational continuity and parts interchangeability. The level of maintenance to be performed on site should be decided early in the design process. This level will determine the type and quantity of spare parts to be purchased and the level of maintenance and repair functions required.

A staff of qualified maintenance personnel is necessary for the assessment system to continue performing at the level established during the system design. Ideally, an experienced maintenance staff should be knowledgeable about the system configuration, subsystem component location, cable routing, installation, checkout, and troubleshooting procedures.

A maintenance log of all equipment repairs and adjustments should be kept to provide a historical record of the actions taken to correct specific problems. Maintenance trends can be established to identify recurring problems and corrective actions to eliminate repetitive occurrences of the same problem. This can substantially reduce repair time and identify equipment performing in a substandard manner.

Adequate maintenance cannot be performed without adequate equipment documentation. Maintenance and operations manuals should be purchased at the time of equipment procurement. Complete documentation should include theory of operation, functional block diagrams, cabling diagrams, schematic diagrams, and parts lists with manufacturers' or commercial equivalent part numbers.

7. OWNER-CONTROLLED AREA SURVEILLANCE

7.1 Overview of Concepts and Definitions

Often, the owner-controlled area of a facility is larger than the section of the facility that is enclosed within protected area (PA) perimeter fencing. (Refer to Figure 80.) Long-distance surveillance technologies are usually employed to survey the owner-controlled area that resides outside the PA perimeter fencing for possible early warning or detection of intruders.

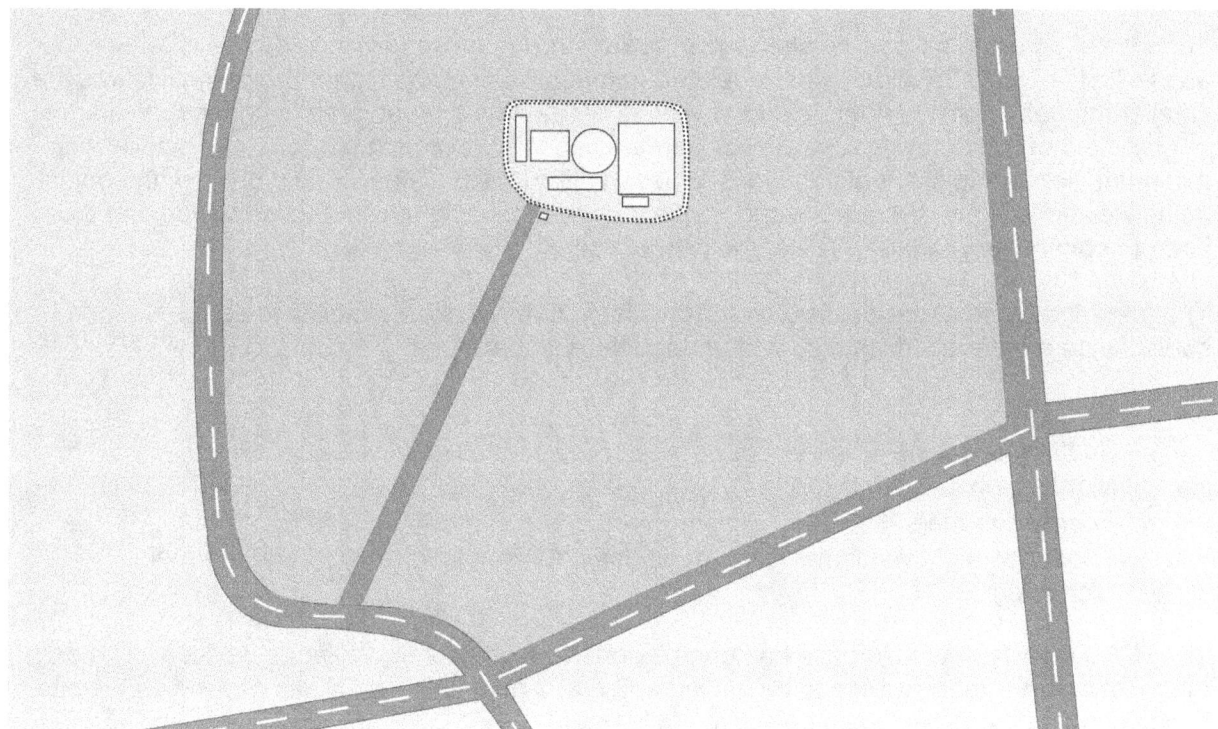

Figure 80: For the example facility, the owner-controlled area, shown in green, is much larger than the security area that is enclosed within perimeter fencing. Real-time surveillance across the owner-controlled area can be challenging, especially given widely varying terrain.

Key elements common to many long-distance surveillance devices that should be understood when considering the purchase of real-time, surveillance-type devices include **magnification**, **objective lens diameter**, and **field of view**.

7.1.1 Magnification

This element determines the degree to which the object being viewed is enlarged. Two numbers are associated with binoculars, for example, 7×50. The "7" represents the amount of magnification; the object being viewed will be 7 times larger than if viewed by the unaided eye. The greater the magnification or power, the smaller the field of view and the less bright the image will be. It is recommended that any item over 10× should be mounted on a tripod or similar device. Spotting scopes typically have three numbers: 15–45×80. The first two numbers "15–45" represent the magnification range of the scope, and the third number ("80") relates to the diameter of the objective lens.

7.1.2 Objective Lens Diameter

Objective lenses are the front lenses of binoculars. The diameter of these lenses, given in millimeters, is the second number associated with binoculars and the third number associated with spotting scopes. For example, for the 7×50 binoculars, the "7" is the magnification or power, and the "50" represents the diameter in millimeters of the objective lenses. The diameter of the objective lens determines the light-gathering ability, which equates to brightness and usability of an image being viewed under certain conditions.

7.1.3 Field of View

The size of an area that can be seen while looking through a surveillance device is referred to as the "field of view." Field of view is related to magnification; the higher the magnification, the smaller the field of view. A large field of view is desirable if a broad area is under surveillance. A large field of view provides the viewer with the capability to continually observe activity with minimum viewing adjustments and loss of observation. With a smaller field of view, the object being viewed is likely to move beyond the field of view causing a loss of observation and the need to continuously adjust. The same is true when the user is moving.

The following types of devices are readily available that will allow a facility to perform surveillance at different distances and under different conditions:

- Binoculars
- Spotting scopes
- Night vision devices
- Thermal imagers
- Enhanced night vision devices, which use a fusion of night vision and thermal technologies

The U.S. Department of Defense has done extensive testing of surveillance devices. The exact model of a surveillance device to be implemented at a particular facility will depend on the site conditions and the surveillance objectives to be met.

7.2 Binoculars

7.2.1 Principles of Operation

Binoculars provide the user with the ability to use both eyes to peer through two identical mirror-symmetrical telescopes. These telescopes magnify distant objects up to 10 times. Because both eyes are used, binoculars allow for depth perception, which helps provide a three-dimensional image. Devices that employ only one eye do not provide depth perception.

Binoculars hold the two telescopes such that a hinge between them allows the distance between the two telescopes to be adjusted to the individual user. A wheel allows the user to focus both telescopes similarly by changing the distance between the eyepieces and the objective lenses. Some binoculars are available that allow the user to focus the two telescopes individually.

The quality of a pair of binoculars is predominantly determined by the quality of the lenses used, which affects the cost of a particular device. Precision-ground lenses can provide a clarity that is not available with cheaper lenses. While the human eye can compensate for poor quality glass for short periods of time, the user will likely experience extreme eye fatigue if more than a few minutes of use is required.

Hand-held binoculars are limited to a 10 times magnification because higher magnification, without the use of a tripod or similar mounting device for image stabilization, results in image vibration/destabilization. The vibration/destabilization is a result of the normal physiology of the body, such as heart and breathing rate.

Binoculars with high-quality lenses will actually help the human eye to see better at night than the unaided eye. (High-quality lenses increase the light-gathering ability of the binoculars, which in turn increases the brightness relative to the human eye at the objective.)

7.2.2 Types of Binoculars Available

A wide variety of binoculars are available that support a wide variety of applications. Binoculars that are considered "general purpose" are typically not appropriate for use by a security organization of a large facility. "Long-range observation," "military," and "range finder" types are usually considered for security applications.

7.2.3 Characteristics and Applications

Most long-range observation binoculars provide good clarity for long-term use. One problem in using such binoculars is that below-optimal lighting conditions can significantly degrade the scene viewed.

Military-rated binoculars generally provide good optics and receive extensive testing, including shock tests, to ensure that they will function as intended under most conditions.

Range-finder binoculars can identify the distance between the user and the object being viewed and are the most expensive type of binocular.

7.2.4 Testing

Testing on a weekly basis is recommended to ascertain that the device operates as advertised by the manufacturer. If the device is dropped, the user should test the device as soon as possible.

7.2.5 Maintenance

Most binoculars require daily cleaning of the lens using a cleaner and soft cloth recommended by the manufacturer. Devices that have an on/off switch also require routine checks and replacement of batteries.

7.3 Spotting Scopes

7.3.1 Principles of Operation

A spotting scope is a small type of telescope that can be used during daylight conditions and can be used any time more magnification is needed than binoculars can provide.

Two factors determine the amount of magnification that can be used in a spotting scope: atmospheric conditions and the optical system.

The view through a spotting scope is significantly impacted by heat waves, dust, humidity, glare, wind, and air currents. The greater the magnification of a device, the greater the degradation of the performance of the spotting scope during such conditions. High altitudes and dry climates favor higher magnification, while low altitudes and wet, humid climates discourage high magnification.

The optical system of a scope is a determining factor in what and how well the user will be able to see. Image quality is directly related to cost. The inexpensive spotting scopes, regardless of size or type, lose image quality as magnification increases. Premium-grade scopes will lose very little image quality as the magnification increases.

The magnification capabilities of spotting scopes range from 10 up to 400 times.

7.3.2 Characteristics and Applications

A quality spotting scope used under good atmospheric conditions can provide high-quality scenes of objects extremely far away. Because of the sizes of lenses needed for the higher magnifications of a spotting scope, these devices can become quite large and cumbersome. Most spotting scopes must be mounted on a tripod or other device to keep them still enough to use.

The higher magnifications and greater distances create a much smaller field of view for the user. Adequate lighting at great distances can also hinder the usability of a scope.

7.3.3 Testing

Testing on a weekly basis is recommended to ascertain that the device operates as advertised by the manufacturer. If the device is dropped, the user should test the device as soon as possible.

7.3.4 Maintenance

Spotting scopes require daily cleaning of the lens using a recommended cleaner and soft cloth.

7.4 Night Vision Devices

7.4.1 Principles of Operation

Night vision devices (also referred to as night vision goggles) operate by amplifying available light. Such devices can take small amounts of light, such as starlight, moonlight, or area-related ambient light, and convert the light energy into electrical energy. Electrons pass through a thin disk containing over 10 million channels. As the electrons pass through these channels, they strike the walls, releasing thousands of more electrons. These multiplied electrons bounce off a phosphor screen, which converts them back into photons and enables the user to see an accurate nighttime view.

Until recently, all image-intensified night vision devices had one attribute in common: they all produced a green image output. A few manufacturers are currently marketing what they call a full-color device.

7.4.2 Types of Night Vision Devices Available

Night vision devices are categorized by generations which reflect the level of technology in their development.

- Generation 0: (1950s) These devices required an infrared light source to illuminate the observation area.

- Generation 1: (1960s) This was the era of the "starlight scope." Still in limited use today, such scopes comprise three intensifier tubes connected in series. These are bulky and heavy in comparison to today's technology.

- Generation 2: (1970s) Use of the microchannel plate (MCP) eliminated the need for back-to-back tubes. Size, weight and image quality were greatly enhanced.

- Generation 3: (1970s–1980s) Two major advances in materials led to an increase in detection at greater distances in much darker conditions and a significant increase in operational life span of the night vision device. Gallium arsenide photocathode and the ion barrier film on the MCP enabled these advancements. I2 (I-squared—Image Intensification) is a common name for the Generation 3 and 4 technologies.

- Generation 4: (1998) Filmless technology with no ion barrier or protective coating on the MCP0 created a 20-percent increase in performance. Unfortunately, the tubes showed immediate degradation since the protective film for the photocathode had been removed. This degradation also affected the life span, reducing it below the required 10,000-hour active life requirement within some branches of the military.

Other items to note are the following:

- AN/PVS (Army Navy/Passive Vision Sight) is a designation prefix for night vision equipment in the U.S. Department of Defense (DOD).

- Variable gain is an improvement capability that enhances the goggles' capability in variable lighting conditions. It provides the user with a rheostat to limit the amount of incoming light.

7.4.3 Characteristics and Applications

Night vision devices can provide valuable images during night-time hours that could never be seen without such a device. Using a night vision device, the distinct image of a moving human can be captured from a distance of 1,000 to over 2,000 meters (3280 to over 6560 feet).

One weakness in the use of night vision devices is the lack of color in the images. Acceptable clarity because of a lack of depth perception in some models can be a concern. Furthermore, a would-be adversary could use particular camouflage that would prevent the user of a night vision device from spotting the adversary. The field of view that the device provides should also be evaluated and should be consistent and appropriate for the type of surveillance and observation required.

7.4.4 Testing

Testing on a weekly basis is recommended to ascertain that the device operates as advertised by the manufacturer. If the device is dropped, the user should test the device as soon as possible.

7.4.5 Maintenance

Most night vision devices require daily cleaning of the lens with a cleaner and soft cloth recommended by the manufacturer. Batteries need to be routinely checked and replaced.

7.5 Thermal Imager Equipment

7.5.1 Principles of Operation

Unlike a night vision device, a thermal imager needs *no* light to operate. Thermal imager devices can detect (see) heat and the movement of heat. Thermal imagers work in the 3 micron to 30+ micron wavelength of the infrared spectrum. This wavelength can be termed as thermal infrared. Thermal infrared is emitted by an object, not reflected off of an object. This emission is picked up by the special lens of thermal vision equipment, focused, and then scanned by a phased array of infrared-detector elements. The detector elements create a detailed temperature pattern called a **thermogram**, which is translated into electric impulses and then translated into data that can be displayed on the user screen.

7.5.2 Types of Thermal Vision Equipment Available

Thermal imagers are available in vehicle- or stationary-mounted, hand-held, or weapon-sight packages. Various models can provide general all-weather use under all light or no light conditions. Some imagers can perform well in fog, smoke, rain, or snow conditions. Thermal imagers have two ways of indicating the presence of intruders: "white hot" means that warmer objects in the picture are shown in white and "black hot" means that warmer objects are shown in black.

Sources of heat from objects that are not of interest to the user can be distracting and may result in the need for investigation and possibly response by members of the security organization.

7.5.3 Characteristics and Applications

Thermal imagers have the ability to operate successfully under many types of conditions in which other surveillance devices are of no value (e.g., all levels of light, smoke, fog, rain, or snow). One of the biggest advantages is that camouflage is not effective if used by adversaries to hide from thermal imagers.

Thermal imagers can weigh several pounds, making them too heavy and cumbersome to carry for prolonged periods. Thermal imagers are more suited to permanent installations than temporary setup. The field of view that the device provides should also be evaluated and should be consistent and appropriate for the specific field application (e.g., thermal rifle optic should not be used for wide-ranging area surveillance as the field of view provided is small). Thermal imagers are susceptible to defeat through shielding the thermal signature of objects within the field of view.

7.5.4 Testing

Testing on a weekly basis and under different environmental conditions is recommended to ascertain that the device operates as advertised by the manufacturer. If the device is dropped, the user should test the device as soon as possible.

7.5.5 Maintenance

Most thermal imager devices require daily cleaning of the lens using a cleaner and soft cloth recommended by the manufacturer. Batteries should be routinely checked and replaced.

7.6 Enhanced Night Vision (Fusion)

7.6.1 Principles of Operation

Enhanced night vision (ENV) devices operate through the **fusion** of the I^2 (Image Intensifier) tube and an infrared micro-bolometer, designed into a compact monocular design. The combination, via overlay, of these two technologies greatly expands the capabilities of a night vision or thermal imager device alone.

7.6.2 Types of Enhanced Night Vision Equipment Available

ENV devices are available as a hand-held observation tool or as a rifle optic. Some recent models allow the user to determine the percentage of night vision use and thermal imager use to enhance viewing capabilities under particular circumstances. (Refer to Figure 81.)

Sources of heat from objects that are not of interest to the user can be distracting and may result in the need for investigation and possibly response by members of the security organization.

7.6.3 Characteristics and Applications

This fusion technology, a combination of thermal imager and night vision technologies, has the ability to operate successfully under many adverse conditions, excelling in poor-visibility environments that would impede other devices.

7.6.4 Installation Criteria

Installers should follow the manufacturer's recommendations.

7.6.5 Testing

Testing on a weekly basis and under different environmental conditions is recommended to ascertain that the device operates as advertised by the manufacturer. If the device is dropped, the user should test the device as soon as possible.

Figure 81 These drawings illustrate the use of an enhanced night vision device, or fusion technology. The top left drawing shows such a device set for 100 percent I² (Image Intensification) use. The top right drawing shows it set for about 50 percent I² and 50 percent infrared use. The bottom drawing shows it set for 100 percent infrared use. Obviously, different situations would require different settings to gain optimal information.

7.6.6 Maintenance

Fusion devices require daily cleaning of the lens with a cleaner and soft cloth recommended by the manufacturer. Batteries should be routinely checked and replaced.

8. ALARM COMMUNICATION AND DISPLAY

8.1 Principles of Operation

8.1.1 Objectives and Goals of the Alarm Communication and Display System

The primary objective of a physical security alarm system is to communicate alarm events received from sensors to a human operator. (Refer to **Error! Reference source not found.Error! Reference source not found.**.) For a system to be effective, it must be capable of operating under all conditions and continuing to operate even when individual components or primary data communication paths fail, which is why redundant data communications capabilities are important. The goals of the system are to report alarms in a timely manner, never lose alarms, and survive any single-point failure with minimal or no degradation in system performance.

8.2 General Guidance for Alarm Communication and Display Systems

In general, alarm communication and display (AC&D) systems should have the qualities described below:

(1) A data communication subsystem should move alarm data in a timely manner. Alarms should be communicated to the alarm station operator within 2 seconds[2] of sensor activation, although less than 1 second is preferred. The data communication subsystem needs to be fast enough to allow the overall alarm system to meet these times. Alarm timing is measured starting at sensor activation and ending at alarm display (visual and audible) to the alarm station operator.

(2) The data communication system should be robust with no single-point failures between field panels and the alarm stations.

- Single-point failures are defined as the loss of the entire or a significant portion of the detection and assessment system capability through the loss of any single **critical component/communications link**.

- A **critical component/communications link** is one that, if failed, the detection and assessment capability of the security system would be completely or significantly degraded, thus requiring security force personnel to replace the functions of detection and assessment to maintain security system effectiveness. **Intrusion detection and assessment systems should have no single component, link, or location that, if failed, would degrade the continued detection and assessment capability of the remainder of the system.**

(3) The system should be designed with redundant data communication links. Redundancy is recommended between sensors and field panels. "Redundancy" refers to the ability of the system to complete the same task through multiple means.

---

[2] Two seconds is a standard time for many security applications. Refer to the section on alarm timing.

(4)   The alarm system and associated data communications systems should not lose any intrusion or entry control alarms, events, tamper alarms, state-of-health messages, or any other alarm necessary to meet the level of protection of the system design.

(5)   The system should provide automatic notification to the alarm station operator when components of the intrusion detection and assessment system fail. This state-of-health monitoring should be continuous, with any faults being reported to the alarm station operators within 2 seconds, although a 1-second notification time is preferred. Also, the system should notify alarm station operators when components return to proper operation.

(6)   The data communication system should be highly reliable, with minimal downtime for repair.

(7)   The overall system should continue to operate in the event of a component failure caused by destruction or tampering at a single physical location.

(8)   The system should be designed to provide automatic "failover" to the greatest extent practicable, enabling continued detection and assessment when single components, communication links, or locations fail. "Failover" refers to the ability to switch to another resource if one component fails (e.g., the central processing unit, or CPU, for the central alarm station (CAS) fails over to the CPU for the secondary alarm station (SAS)).

(9)   Upon restoration of failed equipment or data communication links, the system should return to its original state within 30 seconds. No alarms should be lost while the system is returning to its original state.

(10)   Upon system restoration, the system should be capable of sustaining another failure at a single physical location and still maintain the capability to perform as designed.

8.2.1 Line Supervision/External Connections

In a high-security environment, all computer security networks are dedicated solely to security operations. The computer security network should not be connected to the Internet or other local or wide-area networks not related to security. When communicating between remote sites,[3] all data communications should be encrypted to the level corresponding to the sensitive unclassified or classified nature of the data communications. All system components should be protected against tampering and unauthorized manipulation. Detection, assessment, and access control devices that communicate using radiofrequency (RF) communications are not recommended for use in high-security applications as RF signals can be interrupted or jammed and thus inhibit equipment performance.

In good alarm communication systems, failures of and tampering with critical components associated with alarm detection, transmission, and annunciation are continuously detected and reported as events to the alarm station console. (For timing issues, refer to Section 8.2.2.2, "Alarm Timing.") The data communications system supports line supervision as detailed in Table 9.

---

[3]   Remote sites are sites that, because of their location, are not part of the internal data communications systems. For example, another location must use public lines because dedicated lines are not available.

To protect the system from unauthorized manipulation (e.g., hackers or others having malevolent intent), a security local area network (LAN), if used, should not be connected to any external computer network. The security system should operate on a stand-alone LAN that does not require connections to any external network. However, for associated entry control systems, the stand-alone system could allow (through firewalls and other information technology equipment) temporary or one-way connections to external networks to import badge information encrypted to the line supervision Class A standard. Detection, assessment, and access control devices that communicate using RF communications are not recommended for use in high-security applications as RF signals can be interrupted or jammed and thus inhibit equipment performance. System developers need to protect and secure communication equipment and media against unauthorized access. Equipment cabinets and field hardware containing electronic equipment or cable patch panels should employ tamper alarms to achieve this function.

8.2.2 Alarm Handling

8.2.2.1 Alarm Display Information

Current AC&D systems use both text and graphic displays to uniquely identify alarms to an alarm station operator. Many types of information can be displayed, but some of the more important include the following:

- Mode of any sensor. Mode refers to the status of a sensor. Is the device being monitored or ignored? Is the sensor on line or off line? Other standard terminology includes "access" or "bypass," which means the same as "offline."

- Status or state of an event. Alarm status (state) can include alarming (detecting) or secure (no detection).

- Time of events.

- Location of events.

At a minimum, the alarm station console should display the following information for alarm events:

- Visual and audible annunciation of the alarm

- Alarm type (intrusion, duress, tamper, etc.)

- Alarm location

- Time of alarm

- Site-specific sensor identifier (a minimum of 64 characters is desirable)

- Sensor state (alarm, secure, tamper)

- Sensor mode (access, secure, offline)

Configuring the system to meet the needs of the site or facility is important. Flexibility is desired; for example, users should be able to configure the system to direct a particular alarm, tamper, or system event to the desired console. The system should be able to direct an event to more than one console (for example, sending all intrusion alarms to both the CAS and the SAS consoles). Administrators should be able to direct alarm events to consoles based on their location, priority, or sensor type. Figure 82 shows an example of an alarm station console display.

Text and graphical displays need to be synchronized. Operations performed on a text display should, as needed, cause the graphics display to update. Operations initiated through the graphics display should, as needed, cause the text display to update.

A good AC&D system needs synchronization of all internal "clocks" of all computers and subsystems to a single time base. It is recommended that the system utilize the network time protocol to perform this synchronization. System administrators should ensure that the system is synchronized and configured to display time as the security force requires (usually in the local time zone or universal time).

8.2.2.2   Alarm Timing

Intrusion detection system (IDS) alarms need to be timely so that alarm station operators can assess the alarms and dispatch response forces, as appropriate. In general, alarms, tampers, and system faults need to be annunciated in a timeframe that is a small fraction of the total response time. The IDS timing should be maintained at all times, regardless of other activities happening on the system.

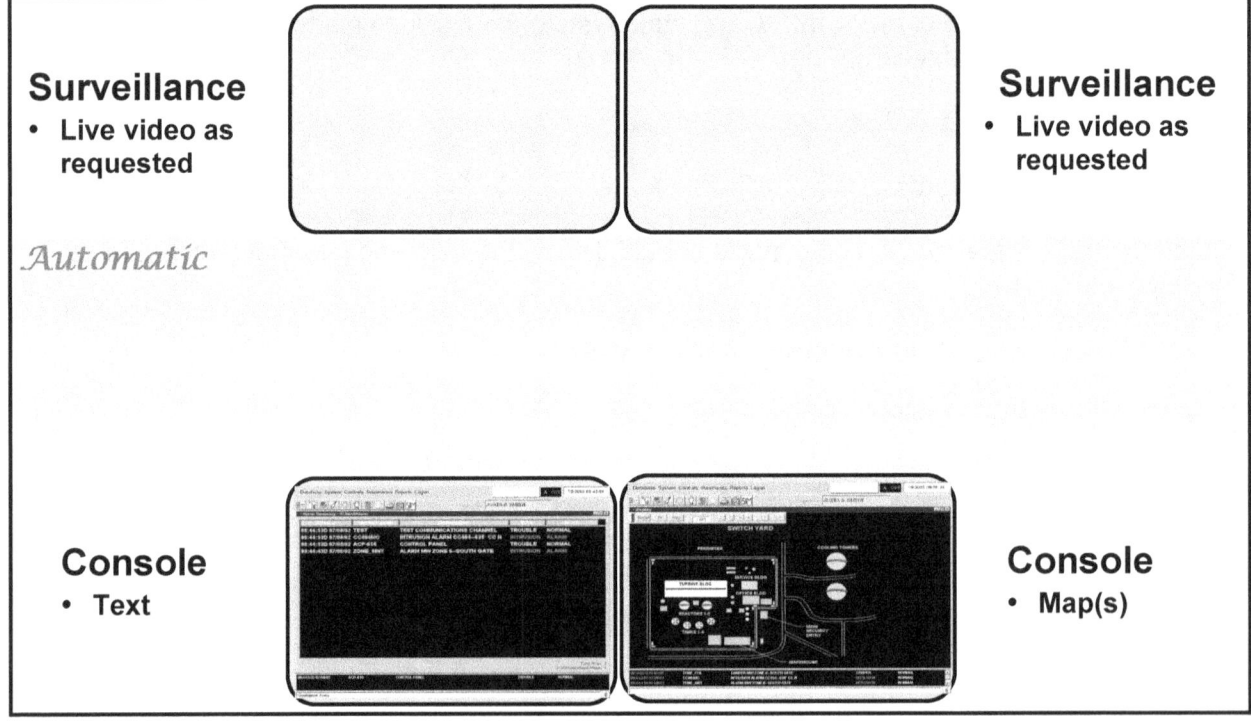

Figure 82: An optimal arrangement of AC&D monitors.

"Time to report" is measured starting from the activation of the sensor, tamper, or component failure and ending when the alarm displays to a human operator in an alarm station. For most high-security system applications, alarms should be reported within 1 to 2 seconds of sensor activation.

In malevolent or attack situations, many IDS alarms will be generated simultaneously. The system needs to handle malevolent situations without failing and without losing any of the multiple alarms. The system should be able to handle up to eight simultaneous alarms (i.e., eight alarms in 2 seconds) and should be able to count and display the number of unacknowledged alarms within 2 seconds of receipt of those alarms. The system should display the first alarm received and indicate the number of remaining alarms. In situations of multiple alarms, tampers, or system faults, the system should not fail or lose any alarm events.

The system should be designed to handle an increased frequency of alarms and system event notifications, without prohibiting the alarm station operator from performing alarm acknowledgment activities on the console. When multiple alarms inundate the alarm console display, the number of unacknowledged system events, alarms, or tampers should not prohibit the proper display of subsequent events.

8.2.2.3 Alarm Priorities

Some IDS events, such as intrusion alarms and tamper alarms, are considered to be higher priority than other system events. In some cases, alarms received from specific intrusion sensors may be considered higher priority than others. The system administrator should assign priority levels to system events within the system, consistent with the site protective strategy and the objectives of the physical protection program. The system software should not make assumptions or place constraints on the priority assigned to a particular event.

The system needs the ability to prioritize the display and handling of system events that include, but are not limited to:

- Intrusion alarms from interior or exterior sensors
- Equipment tampers
- System faults
- Data communication system faults, failures, or tampers
- Alternating current power loss
- Access control alarms and events
- Low battery alarms
- Duress alarms

Prioritization of system events is a site-specific decision that should be made after consideration of the elements of the physical protection program, the site's protective strategy, and the geography of the site. A good AC&D system should allow events to be assigned a priority based on location or event type, but exact priority assignments are site decisions.

The system should automatically handle organization of incoming events. Normally, system events are displayed in order of priority and then by time of arrival. Administrators also need the capability to assign the same priority to different events (all alarms and tampers may have the same priority, for example). Also the system should allow alarm station operators the flexibility to handle any event out of priority order.

8.2.2.4 Alarm Logging and Reporting

Alarm logging lets the system maintain a historical record of alarms so that alarm station operators, system administrators, or other authorized personnel can access a list or log of all system events and activity. The alarm control and display system should have the capability to export system log files into commonly used file formats for reporting or analysis. The alarm log should provide a record of all alarm events and all system events.

Other useful information for the alarm log would include the following:

- Time of operator acknowledgment
- Time of operator cause assessment (completion of operator action)
- A "tagging" ability so that the operator can tag an alarm event with its cause

Using the log, the system should be capable of producing reports of alarm activity and providing the numbers and types of alarms based on their cause or location. The following are examples of reports that are beneficial in evaluating the continued health and maintenance needs of the system:

- Nuisance and false alarm reports detailing the total number of alarms, false alarms and nuisance alarms (based on cause) per sensor zone (location), for a user-selectable time period.

- Sensor grouping report based on type or location. In particular, reports are useful for interior sensors and exterior sensors.

To avoid tampering by an insider, the system should prevent alarm station operators from deleting saved events. In addition, the system should be designed or programmed to ensure that an alarm station operator or system administrator cannot change the status of a detection point or deactivate a locking or access control device without the knowledge and concurrence of the alarm station operator in the other alarm station.

The system can aid alarm station operators and administrators by having several automated features and capabilities. For example, good AC&D systems provide capabilities for logging events and saving them; then, when the log is full, the system should provide for an archive capability to save the information in a long-term storage medium. If the alarm reporting log or storage space (such as a hard disk) becomes filled to capacity, the system should not fail or lose information. A full system log should not cause the system to fail or degrade its performance. System capabilities may include the following:

- Automatic notification (such as an e-mail to the administrator) when the log space becomes limited. The "Alarm Log Full" report can be generated at helpful intervals, such as 80 percent full, 90 percent full, and 95 percent full. Users can archive the log files before any information is lost or new information cannot be added.

- An archive capability to move the log information to removable media, such as CDs or DVDs.

8.2.3 Closed Circuit Television Guidance

Intrusion detection and assessment systems should interface with the alarm station console to provide automatic, integrated operation enabling a closed circuit television (CCTV) assessment system to automatically display video images in the area of the detected activity upon receipt of an alarm. The system should also enable manual switching of assessment assets in response to the alarm station operators' commands in order to provide increased capabilities for assessment.

When evaluating candidate systems, physical security analysts can use the following guidance to help determine the suitability of the CCTV system. The important elements that system purchasers or maintenance personnel may consider include the following:

- The ability to display live video from any camera and the ability to record video from any camera. In addition to the automatic, integrated operation capability, the system should enable an alarm station operator to manually control the switching of camera displays. The system should automatically record video before the alarm event and upon receipt of an IDS alarm and should continue to record for a short time after the alarm is received, thus enabling video recall for further assessment. Also, the system should be capable of archiving past events on a removable medium.

- A minimum equipment set, including four CCTV monitors, two primary assessment monitors (live and recorded), and two surveillance monitors for the operators to view video scenes generated by the site cameras.

- Though a minimum equipment set is suggested, the system should be designed to accommodate the site-specific needs consistent with system configuration and overall size.

8.2.3.1   Assessment Monitors

Monitors dedicated for assessment are under the control of the AC&D system. These monitors play the live and recorded video from an alarm sector when the IDS detects activity within the zone of detection. If no alarm event is being handled, the monitors are either blank or show the highest priority unhandled alarm available to the AC&D system.

The monitors are normally blank unless an alarm is being assessed or the alarm station operator has requested a particular view. When a view is selected by the operator, either for the purpose of alarm assessment or status determination, one of the primary CCTV monitors automatically displays live video coverage from the CCTV camera covering the area requested. If the view contains an alarm, the other primary monitor displays the alarm scene coverage that was recorded automatically on the recorder when the alarm occurred. In most cases, recorded alarm video is "looped" to repeatedly show the few seconds of video before and after the alarm event. If the scene does not contain an alarm, no recorded view is present.

The following scenario describes how a system could automatically process alarm video with involvement from an operator:

(1) When two or more alarms are awaiting assessment, the monitors should show the highest priority live and recorded video scenes, unless the alarm station operator has chosen to view a lower priority alarm. Upon completion of an assessment by the alarm station operator, the system should automatically display the highest priority remaining alarm's video (live and recorded).

(2) At all other times these monitors may be blank. The system should not replace the primary monitor scenes until the operator either ends the assessment or selects a new alarm.

(3) Alarm station operators may view specific alarm sector video by selecting appropriate sectors or sensors on the AC&D console. The AC&D system then displays the live video for that sector. Thus, the assessment monitors are under control of the AC&D system, while the AC&D system is under the control of the alarm station operator.

(4) In some instances two additional assessment monitors might be required for alarm zones that use two video cameras for full video coverage of the zone. In most cases, however, a proper sector design requires only a single camera for assessment.

8.2.3.2 Surveillance Monitors

The free monitors that are not primary alarm assessment monitors should be available to the alarm station operator for surveillance purposes. Alarm station operators should have the capability to assign a camera from any camera view to show live scenes on these surveillance monitors at any time. If one of these monitors is showing a scene in which an alarm occurs, the live scene coverage for the alarming sector will be duplicated; the live scenes will be displayed on the free monitor and on the assessment monitor. These monitors are under the control of the operator and not the AC&D system.

8.2.3.3 Prealarm and Postalarm Video

In addition to live presentations, the system should interface with a video recording subsystem to provide a record of alarm events and allow alarm station operators to replay the camera images from the alarm. Recording capabilities should be automated and actuated by alarm signals. The alarm station console can be the main place to control the video recording system. The system needs to record image frames at a sufficient speed to capture the cause of the alarm. Display monitors need the capability to view both the current live camera information and the recorded information.

Within the recorded video frame(s), the time, date, and location need to be embedded and displayed in the image for current and future reference.

For assessment video recording, video should be recorded several seconds before and several seconds after (pre- and post-alarm video) an alarm event. A good general range of pre- and post-alarm video clips is 3 to 10 seconds, with about 5 seconds being optimal. The AC&D system and associated video recording systems should allow pre- and post-alarm clip recording times to be adjusted for site conditions.

The alarm station operator needs the ability to display any camera image to a surveillance video monitor at any time. Even if new alarms are occurring, any video being displayed on the assessment monitors should not be removed from the monitor unless commanded by the alarm station operator. However, CCTV systems need the capability to alert the operator or to call the operator's attention to an alarm displayed on a video recorder/monitor. The system should not constrain the operator; for example, the operator may need to select video information associated with another incident out of priority order.

CCTV systems should detect and report tampering and loss of video signals.

8.3 Types of Alarm Communication and Display Systems Available

A variety of manufacturers produce AC&D systems. The systems rely on commercial, off-the-shelf computers and network components but are customized through proprietary software created by the manufacturer. When investigating potential vendors of AC&D systems, security analysts should have already defined their needs for the specific facility. In addition, security requirements and specifications for the AC&D system should be defined in advance to assist in evaluating vendors' offerings.

8.4 Sources of Nuisance Alarms

The AC&D system is not a significant source of nuisance alarms. Low-probability communications faults have been known to introduce nuisance events, but recent changes to the communications equipment used in alarm systems has greatly reduced even this low probability. Remote power sources (e.g., those at field panels) can cause nuisance alarms. The primary source of nuisance alarms comes from sensors and not from the AC&D system.

8.5 Installation Criteria

AC&D systems are somewhat different than sensors in that limited, specific criteria apply to all possible systems. One criterion for the installation of an AC&D system and its communications and power is that these systems must be installed in accordance with the National Electrical Code as is the case with many industrial electrical components.

In general, AC&D vendors have either specific installation criteria or instructions that must be followed. Many vendors allow only fully trained and qualified distributors to install their systems. For other systems, a team of properly trained engineering and technical personnel will be necessary to correctly install a complex AC&D system. AC&D vendors are the best source of information on the criteria necessary for installing their systems.

8.6 Testing

8.6.1 Failover Testing

The system will occasionally fail because of power outages or component failure. Failover testing involves the ability to simulate these power or mechanical failures and then check the system to ensure that it responds accurately to system failures. Failover testing involves the following actions:

- Identifying all critical components through the use of fault tree analysis (refer to Figure 85) or similar techniques. The fault tree allows analysts to explore all the components of a system and define all the events (single or in combinations) that could cause a system failure. Using standard logic symbols (Figure 84), analysts build system maps to show all the undesired system states and the components that could contribute to a system failure. The fault tree analysis technique has been used in safety and risk engineering for the determination of safety hazards.

- Identifying failure causes for all components discovered during fault tree analysis (e.g., servers, workstations, switchers, routers, other data communications devices, and power sources).

- Causing failures in various components of a system.

- Verifying that the system continues to communicate data and display alarm information before, during, and after the failure.

- Verifying that the system restores to its original state once the failure is removed and that alarms continue to be captured, communicated, and displayed during the restoration.

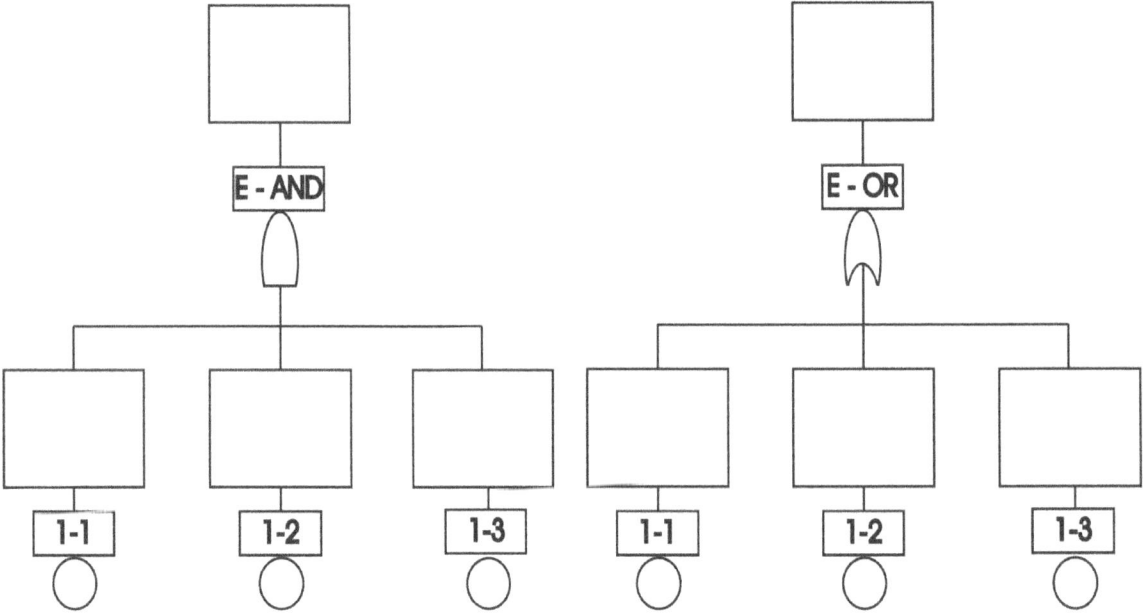

Figure 83: Fault trees allow analysts to map events and their potential causes. (Refer to the next figure for explanations of symbols.)

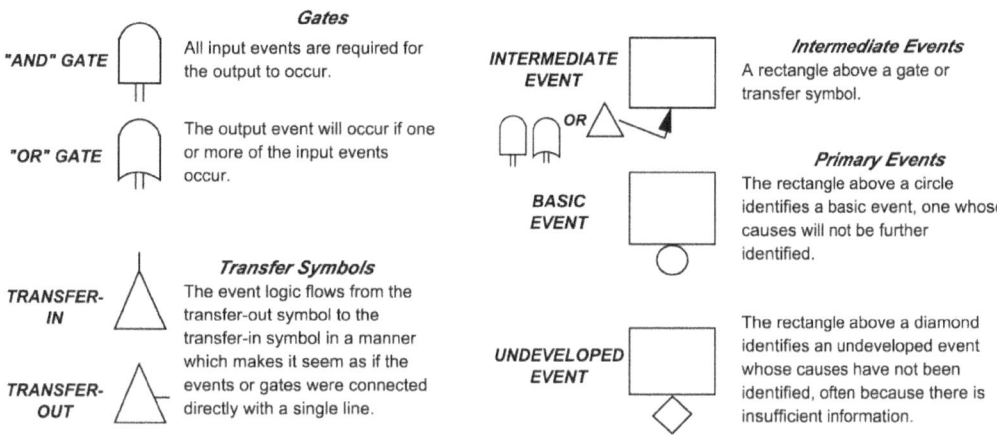

Figure 84: Symbols used in the graphical representation of a fault tree.

8.6.2 Acceptance Testing

Acceptance testing is typically performed at installation and verifies that the system meets all specifications. Acceptance testing typically consists of tests that challenge and demonstrate the performance criteria associated with the manufacturer's specifications. Acceptance tests should be conducted anytime the system software changes or major hardware additions are installed.

8.6.3 Operability Testing

Operability tests should be conducted during the normal operation of the system. In most cases AC&D operability can be tested by performing sensor and video operability tests and observing that generated events are displayed properly on the AC&D console.[4]

Table 9 shows an example of AC&D operability testing.

8.6.4 Performance Testing

Performance testing of the AC&D system is typically done as part of the acceptance test process. Those requirements that have a measurable performance metric are tested appropriately (timing tests, for example, are measured with external time measurement devices). In general, the AC&D system performance changes only when hardware or software changes are made. There is little degradation or component drift associated with a good AC&D system. This is a distinct difference between the AC&D system and its associated sensors. A semi-annual performance test is recommended. A performance test should also be conducted after major hardware or software installations.

---

[4] Many sites conduct ongoing operability tests (weekly or monthly) throughout the year on a schedule that ensures that all alarm points are covered at least once per year. Other effective schedules are possible.

Table 9: Operations Testing Matrix

Required Feature	Test To Ensure Operability
Time, date, and location are embedded in the image of the recorded video.	Alarm station operator inspects console display of the recorded video for the time, date, and location.
Video should be recorded several seconds before and several seconds after (pre- and post-alarm video) an alarm event.	While an alarm station operator works at the console, a security officer should perform a test of the IDS at the desired location. The alarm station operator inspects console display of the recorded video for the site-specified number of seconds for pre- and post-alarm video.
Display any camera image to a surveillance video monitor at any time.	Alarm station operator ensures that all cameras respond to commands to display on the system in the order chosen by the operator.
CCTV systems should detect and report tampering and loss of video signals.	While an alarm station operator works at the console, a security officer should perform an intentional tamper with the CCTV apparatus to cause an alarm at the console. To test for loss of video signals, the security officer in the field should turn off the camera to ensure that the alarm station operator receives a system notification of the signal loss.

8.7 Maintenance

A strong, proactive maintenance program should be in place to maintain the AC&D subsystem. Manufacturers of AC&D systems have specific recommendations on the maintenance requirements. The appropriate maintenance personnel should be on site or on call to meet required "mean time to repair" requirements specified by the vendors or be available to meet the site-specific requirements, which may vary depending on the environmental conditions at the site.

9. EMERGENCY AND BACKUP POWER

9.1 Principles of Operation

The primary power source for critical security systems and components for a limited area should come directly from the normal onsite power distribution system or directly from the public utility. Because of the potential for a loss of power, whether caused by a natural phenomenon such as weather or by a malicious act committed by an adversary, a facility should have reliable backup and/or emergency power sources on site that will provide power when needed.

All critical security systems and components, including the following, should have an emergency/backup power capability in case of the loss of primary power if they are required to function as intended to maintain a high physical protection standard:

- Intrusion detection equipment
- Assessment equipment (closed circuit television systems)
- Illumination equipment (lights required for assessment cameras)
- Automated access control systems
- Alarm monitoring systems
- Nonportable communications equipment

These power sources should contain an automatic switching capability to the auxiliary source of power (battery and/or generator) that will function immediately without causing false alarms and without causing a loss of security system function or data. (See Figure 85.)

Figure 85: A facility must have emergency and/or backup power supplies so that in the event of a loss of power, critical parts of the security system will continue to operate without loss of key information or functionality.

It is important that the alarm stations be designed such that they will automatically receive an alarm indicating any of the following conditions:

- The primary power source has failed.
- The facility has transferred to an emergency/backup power source.
- The emergency/backup power source has failed.

In the design and selection of an emergency/backup power source, consideration should be given to site-specific conditions for the capability to restore primary power. The capability of the emergency/backup power source to sustain security system operations should be based on the timeframe to restore primary power as derived through a site specific analysis.

9.2 Types of Emergency and Backup Power Supplies

While a wide variety of emergency/backup power mechanisms are available, the three primary categories that are used by most facilities are the following:

- Uninterruptible power supplies (UPSs)
- Batteries
- Engine generators (EGs)

9.2.1 Uninterruptible Power Supplies

UPSs are typically used to supply an uninterrupted source of power to important instrumentation and control systems enabling continuous operation during the loss of normal power without loss of system or component functionality. They are also used to provide continuous, quality power for systems sensitive to disturbances occurring in an electrical power distribution system caused by switching, faults, or power transfer. UPS designs may include various combinations of batteries, rectifier/charger, battery transfer, bypass switch, and inverter. Most UPS systems are placed between the primary power source and the component it protects, effectively funneling all power at all times while keeping its own batteries charged.

9.2.2 Battery Systems

A battery system includes all switchgear and distribution equipment necessary to provide quality voltage and current as required by the connected load. The batteries are normally in full float operation where they are connected in parallel with a charger and the load, and where the charger supplies the normal direct current (dc) load plus any self-discharge or charging current, or both, required by the battery.

9.2.3 Engine Generators

An EG is a device that converts mechanical energy to electrical energy, generally by using electromagnetic induction. For most industrial applications, a diesel engine supplies the mechanical energy. An EG, when properly sized and designed, will provide reliable electrical power to the intended load equipment for the required amount of time.

9.3 Installation Criteria

Directions and recommendations provided by the manufacturer should be followed as appropriate. Emergency/backup power systems and key components, including batteries, EGs, fuel tanks, and switch gear, should be physically located where they will be protected; they should either be under continual surveillance or contained in a locked enclosure with an intrusion detection system installed to protect against tampering and unauthorized access.

Nonrechargeable batteries should be replaced whenever their voltage drops 20 percent below the rated voltage or based on the manufacturer's recommendation. An alarm signal should alert the alarm stations of this condition.

An EG must be capable of providing power compatible with the equipment it sources. The EG should have adequate capacity and rating for all loads to be operated simultaneously. The emergency/backup power system should start automatically (or be brought on line) upon loss of primary power. The transfer of power from battery to EG power should occur at a set point before the battery power declines in voltage below the levels needed to continue the operation of assigned equipment. This transfer is typically set to occur prior to battery voltage falling below 80 to 90 percent of normal in the event the generator initially fails to start.

9.4 Testing

A regular program of testing emergency/backup power sources is a key to keeping them in optimal operating order. Three types of testing need to be performed at different times in the life of a power system: acceptance testing, performance testing, and operability testing.

9.4.1 Acceptance Testing

Acceptance testing is the process that a site must go through after an installer has completed the installation of the system but before the system is used operationally.

Acceptance testing consists of two parts:

(1)   The newly installed components should be thoroughly inspected to ensure that the system has been installed according to manufacturer's specifications and the detailed engineering drawings for the site-specific design/installation. A facilities engineer will need to ascertain that commonly accepted practices for electrical, mechanical, and plumbing work have been followed, including safety requirements.

(2)   A complete and positive performance test should be performed (see next section).

9.4.2 Performance Testing

Performance testing is the periodic test of the full range of expected capabilities of a particular emergency/backup power system. Most important, the emergency/backup power system should be demonstrated to fulfill the following purposes:

- The system receives instant notification when primary power has failed.
- The system becomes fully operational within the timeframe for which it was designed.
- The system provides the power voltage for which it was designed.

- The system continues to operate for the length of time for which it was designed.
- The system can reliably return to normal power when appropriate.

This complete test should be accomplished at least once every 12 to 18 months, or whenever there has been a change to the system, including the rerouting of power sources, movement of equipment, or an upgrade of features or components.

9.4.2.1 Engine Generator Testing

An integrated system test should be performed at least once a year to demonstrate proper operation during a loss of primary power. This series of tests demonstrates the ability of the EG unit and associated switch gear to perform their intended function under simulated accident conditions. In addition to the steps performed during the monthly test outlined in the next section, perform the following steps should be performed:

(1)   Verify that the buses that will be powered by the EG deenergize and loads shed, as required.

(2)   Verify that the EG starts automatically and attains voltage and frequency within prescribed limits and time.

(3)   Verify the proper loading sequence and that voltage and frequency are within the manufacturer's specifications for connected loads for both transient and steady-state conditions.

(4)   Run the EG for at least 8 hours after temperature equilibrium is reached.

(5)   Demonstrate that the EG operates at its continuous rating and, if installation permits, efficiency rating (also called the load power factor). If the connected load is above the continuous rating but within the short-time rating, the unit should be operated at the short-time overload value for no longer than the time specified for the short-time rating.

(6)   After the test is complete, shut the unit down and within 5 minutes demonstrate the ability of the unit to perform a hot restart and load to its continuous rating.

9.4.2.2 Uninterruptible Power Supply Testing

A UPS in its standby or normal operating mode may not demonstrate many of the various features that may be required to function during emergency conditions, which include, but are not limited to, a loss-of-power or equipment failure.

The tests outlined below are recommended UPS tests and, depending on the design of the UPS system, should be performed according to manufacturer's recommendations or on at least an 18-month interval:

- Light-Load Test—Operate controls and instruments for stability and values of voltage and frequency.

- Synchronization Test—Measure the rate of frequency change during synchronization and UPS voltage during transfer (when an alternate source is part of the design).

- Alternating Current Input Failure Test—Verify that transfer to dc source occurs as designed.

- Alternating Current Input Return Test—Verify that system performs stable return to normal source.

- Transfer Test—Forward and reverse (UPS systems using static bypass switches).

- Rated Full-Load Test—Check that system meets connected or rated load-carrying capability for the required duration for extremes of alternating current (ac) and dc input voltage.

- Output-Voltage Balance Test—Measure phase angle and voltage to meet specifications for balanced and unbalanced conditions.

- Harmonic-Components Test—Measure harmonic content in the output voltage for linear and nonlinear load conditions.

9.4.2.3 Testing Batteries

The only real measure of a battery's capacity and capability to provide power to its required load is derived from the performance of two different tests: the performance discharge test and the service test. During the performance of these tests, the battery will be unavailable for duty because of significant discharge. These tests should be performed only under conditions when the unavailability of the battery is acceptable, or provisions should be made (compensatory measures) for an alternate source to be temporarily connected to the loads for the duration of the test and recharging of the battery.

The performance discharge test indicates the remaining capacity of the battery expressed as a percent of the rated capacity. This provides an indication of the remaining useful life of the battery.

The service test demonstrates the ability of the battery to carry its required load for the required time period. To perform the service test, the load profile for the battery must be known.

9.4.3 Operational Testing

Operational testing is the monthly testing of an emergency/backup power system to confirm that it is functional. Because regular maintenance is critical in many systems, it is highly recommended that this operational testing requirement be performed at the same time as regular maintenance. Because it is sometimes difficult to distinguish between what constitutes periodic testing and what constitutes periodic maintenance, the two have been combined in Section 9.5, "Maintenance."

Table 10 (taken from DOE-STD-3003-2000) provides an example of good practices for surveillance and testing of lead-acid cells, including expected values and testing frequency. Adjustments (increases or decreases) to the intervals in the table should be based on experience and manufacturer's recommendations.

Table 10: Typical Lead-Acid Battery/Cell Surveillance and Tests

Subject (Note 3)	Values (Note 1)	Period
Battery Terminal Voltage (typical 60-cell battery)	(129–130.2) volts PbSb (130.2–135) volts PbCa	Monthly
Electrolyte Level	Between fill lines (distilled water only)	Monthly
Cell Float Voltage (Pilot Cell)	(2.15–2.17) volts PbSb (2.17–2.25) volts PbCa	Monthly
Cell Float Voltage (All Cells)	(2.15–2.17) volts PbSb (2.17–2.25) volts PbCa	Quarterly
Specific Gravity (Pilot Cell)	1.215 ± 0.010	Monthly
Specific Gravity (All Cells)	1.215 ± 0.010	Quarterly
Cleanliness and Corrosion Check	(Note 2)	Monthly
Resistance Measurement Cell to Cell and Terminal Connections	< 20% above the value when new or after cleaning or retorquing bolts	Yearly
Battery Capacity Test (Performance Discharge Test)	> 80% of capacity (Note 5)	5 years
Battery Load Test (Service Test)	Design load and duration with adequate voltage	Yearly (Note 4)

Note 1   Values outside these ranges indicate that action is necessary to restore the parameter to within the specified values.

Note 2   Battery cleaning should be done with baking soda and water with clear water rinse—no other cleaning materials should be used.

Note 3   All manufacturer's safety precautions should be observed when working on batteries.

Note 4   Not required where battery is for engine generator only and not other loads. The periodic monthly start of the engine generator is a load test. Load tests should be done in as-found condition.

Note 5   Battery capacity (performance discharge) test is used as a battery age indicator. Initially, a battery should be fully charged, temperature near 25°C, and terminals clean.

9.5 Maintenance

Because of the nature of emergency/backup power supplies, as well as their expected irregular schedule of use, regular maintenance is critical to guaranteeing their full functionality and reliability when called upon when the primary power source is unavailable.

9.5.1 Engine Generators

To avoid unnecessary and premature degradation of the EG, the manufacturer's recommendations should be followed in regard to prelubrication, acceleration, loading, and unloading. All engine starts, for testing purposes, should follow the manufacturer's recommendations for prelubrication and/or other warmup procedures to minimize mechanical stress and wear.

Regular exercising of the EG keeps engine parts lubricated, prevents oxidation of electrical contacts, uses up fuel before it deteriorates, and in general, helps provide reliable engine starting. The generator set should be run at least once a month for a minimum of 1 hour loaded to no less than one-third of the nameplate rating.

The following, at a minimum, should be performed at least once per month, unless otherwise specified by the manufacturer:

- Record the as-found condition and the identity of the test personnel.

- Verify that the EG starts on a simulated loss of offsite power and accelerates to operating rpm in (the required value) seconds. Generator voltage and frequency should be (required value + 10 percent) volts and (60 + 2 percent) hertz within (required value) seconds.

- Verify that the starting system disengages properly.

- Verify that fuel tank levels (fuel storage tank, day tank, and engine tank) are within specifications.

- Remove accumulated water from fuel tanks and oil/water separator.

- Verify that the fuel transfer pump system starts and that fuel is being transferred from the storage tank to the day and engine-mounted tank (if present).

- Verify that the EG can accept loads up to 90 percent of the continuous rating for the duration of the test (time should consider the manufacturer's recommendations). This may be done using normal loads or a load bank, or by synchronizing to the offsite grid if this capability is available. If a load bank is routinely used, a dedicated connection should be provided. The value of 90 percent is used to provide margin to avoid inadvertent overloading.

- Verify the cooling water tank level.

When recording test results, a failure should be recorded if the EG fails to start, accelerate, reach nominal voltage and frequency, or accept the rated load within and for the time required, or if it otherwise fails to satisfy the specified acceptance criteria.

9.5.2 Uninterruptible Power Supplies

The UPS system should be inspected and tested on a frequent basis according to the manufacturer's recommendations. The frequency of the inspections is dictated by the type of service to which the equipment is subjected (e.g., duty cycle, chemicals, dust and heat) and trending. Less frequent activities such as internal cleaning, filter replacement, checking electrical connections for tightness, and calibration of instruments should be done according to the manufacturer's recommendations.

9.5.3 Batteries

Many types of batteries, if allowed to sit without a charger, will internally discharge, often with irreversible cell degradation. For these types of batteries, it is important to maintain proper charging float voltage during standby. Because of inherent differences between cells, float voltage and specific gravity values will vary from cell to cell over time. If cell float voltages and/or specific gravity values are allowed to remain in an unequal condition for extended periods of time, cell sulfation will result. To overcome this problem, a periodic equalizing charge must be applied to equalize cell voltages and specific gravities. The manufacturer's recommendations should be followed in regard to equalizing charge. When performing an equalizing charge, care should be taken to ensure that the charger voltage does not exceed the voltage rating of the load connected during the equalize charge. Batteries are usually rated at a temperature of 25 degrees Celsius. Higher temperatures will improve capacity at high discharge rates but significantly reduce battery life. Lower temperatures have a significant effect in reducing battery capacity. Typical battery types for standby service are lead-acid (calcium (Ca), antimony (Sb)), pure lead (Pb) (generally a "round cell"), or nickel-cadmium (NiCd). Manufacturers will provide necessary information on the care, precautions, charging, and treatment of specific batteries including maintenance during periods of storage.

NRC FORM 335
(12-2010)
NRCMD 3.7

U.S. NUCLEAR REGULATORY COMMISSION

1. REPORT NUMBER (Assigned by NRC, Add Vol., Supp., Rev., and Addendum Numbers, if any.)

BIBLIOGRAPHIC DATA SHEET

(See instructions on the reverse)

NUREG-1959

2. TITLE AND SUBTITLE	3. DATE REPORT PUBLISHED	
Intrusion Detection Systems and Subsystems	MONTH	YEAR
	March	2011
	4. FIN OR GRANT NUMBER	

5. AUTHOR(S)	6. TYPE OF REPORT
Division of Security Policy, Office of Nuclear Security and Incident Response	
	7. PERIOD COVERED (Inclusive Dates)

8. PERFORMING ORGANIZATION - NAME AND ADDRESS (If NRC, provide Division, Office or Region, U.S. Nuclear Regulatory Commission, and mailing address; if contractor, provide name and mailing address.)

Division of Security Policy
Office of Nuclear Security and Incident Response
U.S. Nuclear Regulatory Commission
Washington, DC 20555-0001

9. SPONSORING ORGANIZATION - NAME AND ADDRESS (If NRC, type "Same as above"; if contractor, provide NRC Division, Office or Region, U.S. Nuclear Regulatory Commission, and mailing address.)

Same as above

10. SUPPLEMENTARY NOTES

11. ABSTRACT (200 words or less)

This report provides information relative to designing, installing, testing, maintaining, and monitoring intrusion detection systems (IDSs) and subsystems used for the protection of facilities licensed by the U.S. Nuclear Regulatory Commission. It contains information on the application, use, function, installation, maintenance, and testing parameters for internal and external IDSs and subsystems, including information on communication media, assessment procedures, and monitoring. This information is intended to assist licensees in designing, installing, employing, and maintaining IDSs at their facilities.

12. KEY WORDS/DESCRIPTORS (List words or phrases that will assist researchers in locating the report.)	13. AVAILABILITY STATEMENT
Intrusion Detection Systems and Subsystems.	unlimited
	14. SECURITY CLASSIFICATION
	(This Page) unclassified
	(This Report) unclassified
	15. NUMBER OF PAGES
	16. PRICE

NRC FORM 335 (12-2010)

UNITED STATES
NUCLEAR REGULATORY COMMISSION
WASHINGTON, DC 20555-0001

OFFICIAL BUSINESS

NUREG-1959

Intrusion Detection Systems and Subsystems

March 2011